Reliability Analysis, Safety Assessment and Optimization

Wiley Series in Quality & Reliability Engineering

Dr. Andre V. Kleyner
Series Editor

The Wiley Series in Quality & Reliability Engineering aims to provide a solid educational foundation for both practitioners and researchers in the Q&R field and to expand the reader's knowledge base to include the latest

developments in this field. The series will provide a lasting and positive contribution to the teaching and practice of engineering.

The series coverage will contain, but is not exclusive to,

- Statistical methods
- Physics of failure
- Reliability modeling
- Functional safety
- Six-sigma methods
- Lead-free electronics
- Warranty analysis/management
- Risk and safety analysis

Wiley Series in Quality & Reliability Engineering

Reliability Analysis, Safety Assessment and Optimization: Methods and Applications in Energy Systems and Other Applications
by Enrico Zio, Yan-Fu Li
May 2022

Design for Excellence in Electronics Manufacturing
Cheryl Tulkoff, Greg Caswell
April 2021

Design for Maintainability
by Louis J. Gullo (Editor), Jack Dixon (Editor)
March 2021

Reliability Culture: How Leaders can Create Organizations that Create Reliable Products
by Adam P. Bahret
February 2021

Lead-free Soldering Process Development and Reliability
by Jasbir Bath
(Editor) August 2020

Automotive System Safety: Critical Considerations for Engineering and Effective Management
Joseph D. Miller
February 2020

Prognostics and Health Management: A Practical Approach to Improving System Reliability Using Condition-Based Data
by Douglas Goodman, James P. Hofmeister, Ferenc Szidarovszky
April 2019

Improving Product Reliability and Software Quality: Strategies, Tools, Process and Implementation, 2nd Edition
Mark A. Levin, Ted T. Kalal, Jonathan Rodin
April 2019

Practical Applications of Bayesian Reliability
Yan Liu, Athula I. Abeyratne
April 2019

Dynamic System Reliability: Modeling and Analysis of Dynamic and Dependent Behaviors
Liudong Xing, Gregory Levitin, Chaonan Wang
March 2019

Reliability Engineering and Services
Tongdan Jin
March 2019

Design for Safety
by Louis J. Gullo, Jack
Dixon February 2018

Thermodynamic Degradation Science: Physics of Failure, Accelerated Testing, Fatigue and Reliability by Alec Feinberg October 2016

Next Generation HALT and HASS: Robust Design of Electronics and Systems
by Kirk A. Gray, John J. Paschkewitz May 2016

Reliability and Risk Models: Setting Reliability Requirements, 2nd Edition
by Michael Todinov November 2015

Reliability Analysis, Safety Assessment and Optimization

Methods and Applications in Energy Systems and Other Applications

Enrico Zio
MINES ParisTech/PSL Université, Italy

Yan-Fu Li
Department of Industrial Engineering, Tsinghua University, China

WILEY

This edition first published 2022

Registered Offices
John Wiley & Sons, Inc., 111 River Street, Hoboken, NJ 07030, USA
John Wiley & Sons Ltd, The Atrium, Southern Gate, Chichester, West Sussex, PO19 8SQ, UK

Editorial Office
The Atrium, Southern Gate, Chichester, West Sussex, PO19 8SQ, UK

For details of our global editorial offices, customer services, and more information about Wiley products visit us at www.wiley.com.

Wiley also publishes its books in a variety of electronic formats and by print-on-demand. Some content that appears in standard print versions of this book may not be available in other formats.

Library of Congress Cataloging-in-Publication Data
Names: Zio, Enrico, author. | Li, Yan-Fu, author.
Title: Reliability analysis, safety assessment and optimization: methods and applications in energy systems and other applications / Enrico Zio, Yan-Fu Li.
Description: Hoboken, NJ : John Wiley & Sons, 2022. | Series: Wiley series in quality & reliability engineering | Includes bibliographical references and index.
Identifiers: LCCN 2022009590 (print) | LCCN 2022009591 (ebook) | ISBN 9781119265870 (hardback) | ISBN 9781119265924 (pdf) | ISBN 9781119265863 (epub) | ISBN 9781119265856 (ebook)
Subjects: LCSH: Reliability (Engineering) | Industrial safety.
Classification: LCC TA169 .Z57 2022 (print) | LCC TA169 (ebook) | DDC 620/.00452--dc23/eng/20220316
LC record available at https://lccn.loc.gov/2022009590
LC ebook record available at https://lccn.loc.gov/2022009591

Cover image: © Vithun Khamsong/Getty Images
Cover design by Wiley

Set in 9.5/12.5pt STIXTwoText by Integra Software Services Pvt. Ltd, Pondicherry, India
Printed and bound by CPI Group (UK) Ltd, Croydon, CR0 4YY

C9781119265870_200522

To all the students, collaborators and colleagues, to whom we are forever indebted for all the enriching experience and knowledge that they have shared with us and that have made us grow professionally and personally.

Contents

Series Editor's Foreword

Dr. Andre V. Kleyner

The Wiley Series in Quality & Reliability Engineering aims to provide a solid educational foundation for researchers and practitioners in the field of quality and reliability engineering and to expand the knowledge base by including the latest developments in these disciplines.

The importance of quality and reliability to a system can hardly be disputed. Product failures in the field inevitably lead to losses in the form of repair cost, warranty claims, customer dissatisfaction, product recalls, loss of sale, and in extreme cases, loss of life.

Engineering systems are becoming increasingly complex with added functions and capabilities; however, the reliability requirements remain the same or even growing more stringent. Modeling and simulation methods, such as Monte Carlo simulation, uncertainty analysis, system optimization, Markov analysis and others, have always been important instruments in the toolbox of design, reliability and quality engineers. However, the growing complexity of the engineering systems, with the increasing integration of hardware and software, is making these tools indispensable in today's product development process.

The recent acceleration of the development of new technologies including digitalization, forces the reliability professionals to look for more efficient ways to deliver the products to market quicker while meeting or exceeding the customer expectations of high product reliability. It is important to comprehensively measure the ability of a product to survive in the field. Therefore, modeling and simulation is vital to the assessment of product reliability, including the effect of variance on the expected product life, even before the hardware is built. Variance is present in the design parameters, material properties, use conditions, system interconnects, manufacturing conditions, lot-to-lot variation and many other product inputs, making it difficult to assess. Thus, modeling and simulation may be the only tools to fully evaluate the effect of variance in the early product development phases and to eventually optimize the design.

The book you are about to read has been written by leading experts in the field of reliability modeling, analysis, simulation and optimization. The book covers important topics, such as system reliability assessment, modeling and simulation, multi-state systems, optimization methods and their applications, which are highly critical to meeting the high demands for quality and reliability. Achieving the optimal feasible performance of the system is eventually the final objective in modern product design and manufacturing, and this book rightfully puts a lot of emphasis on the process of optimization.

Paradoxically, despite its evident importance, quality and reliability disciplines are somewhat lacking in today's engineering educational curricula. Only few engineering schools offer degree programs, or even a sufficient set of courses, in quality and reliability methods. The topics of reliability analysis, accelerated testing, reliability modeling and simulation, warranty data analysis, reliability growth programs, reliability design optimization and other aspects of reliability engineering receive very little coverage in today's engineering students curricula. As a result, the majority of the quality and reliability practitioners receive their professional training from colleagues, professional seminars and professional publications. In this respect, this book is intended to contribute to closing this gap and provide additional educational material as a learning opportunity for a wide range of readers from graduate level students to seasoned reliability professionals.

We are confident that this book, as well as this entire book series, will continue Wiley's tradition of excellence in technical publishing and provide a lasting and positive contribution to the teaching and practice of reliability and quality engineering.

Preface

Engineering systems, like process and energy systems, transportation systems, structures like bridges, pipelines, etc., are designed to ensure successful operation throughout the anticipated service lifetime in compliance with given all-around sustainability requirements. This calls for design, operation, and maintenance solutions to achieve the sustainability targets with maximum benefit from system operation. Reliability, availability, maintainability and Safety criteria (RAMS) are among the indicators for measuring system functionality with respect to these intended targets.

Today, modern engineering systems are becoming increasingly complex to meet the high expectations by the public for high functionality, performance, and reliability, and with this, RAMS properties have become further key issues in design, maintenance, and successful commercialization.

With high levels of RAMS being demanded on increasingly complex systems, the reliability assessment and optimization methods and techniques need to be continuously improved and advanced. As a result, many efforts are being made to address various challenges in complex engineering system lifecycle management under the global trend of systems integration. Mathematically and computationally, the reliability assessment and optimization are challenged by various issues related to the uncertain, dynamic, multi-state, non-linear interdependent characteristics of the modern engineering systems and the problem of finding optimal solutions in irregular search spaces characterized by non-linearity, non-convexity, time-dependence and uncertainty.

In the evolving and challenging RAMS engineering context depicted above, this book provides a precise technical view on system reliability methods and their application to engineering systems. The methods are described in detail with respect to their mathematical formulation and their application is illustrated through numerical examples and is discussed with respect to advantages and limitations. Applications to real world cases are given as a contribution to bridging the gap between theory and practice.

The book can serve as a solid theoretical and practical basis for solving reliability assessment and optimization problems regarding systems of different engineering disciplines and for further developing and advancing the methods to address the newly arising challenges as technology evolves.

Reliability engineering is founded on scientific principles and deployed by mathematical tools for analyzing components and systems to guarantee they provide their functions as intended by design.

On the other hand, technological advances continuously bring changes of perspectives, in response to the needs, interests, and priorities of the practical engineering world. As technology advances at a fast pace, the complexity of modern engineered systems increases and so do, at the same time, the requirements for performance, efficiency, and reliability. This brings new challenges that demand continuous developments and advancements in complex system reliability assessment and optimization.

Therefore, system reliability assessment and optimization is inevitably a living field, with solution methodologies continuously evolving through the advancements of mathematics and simulation to follow up the development of new engineering technology and the changes in management perspectives. For this, advancements in the fields of operations research, reliability, and optimization theory and computation continuously improve the methods and techniques for system reliability assessment and optimization and for their application to very large and increasingly complex systems made of a large number of heterogeneous components with many interdependencies under various physical and economic constraints.

Within the ongoing efforts of development and advancement, this book presents an overview of methods for assessing and optimizing system reliability. We address different types of system reliability assessment and optimization problems and the different approaches for their solutions. We consider the development and advancement in the fields of operations research, reliability, and optimization theory to tackle the reliability assessment and optimization of complex systems in different technological domains.

The book is directed to graduate students, researchers and practitioners in the areas of system reliability, availability, maintainability and Safety (RAMS), and it is intended to provide an overview of the state of knowledge of and tools for reliability assessment and system optimization. It is organized in three parts to introduce fundamentals, and illustrate methods and applications.

The first part reviews the concepts, definitions and metrics of reliability assessment and the formulations of different types of reliability optimization problems depending on the nature of the decision variables and considering redundancy allocation and maintenance and testing policies. Plenty of numerical examples are provided to accompany the understanding of the theoretical concepts and methods.

The second part covers multi-state system (MSS) modeling and reliability evaluation, Markov processes, Monte Carlo simulation (MCS), and uncertainty treatment under poor knowledge. The reviewed methods range from piecewise-deterministic Markov processes (PDMPs) to belief functions.

The third part of the book is devoted to system reliability optimization. In general terms, system reliability optimization involves defining the decision variables, the constraints and the single or multiple objective functions that describe the system reliability performance and involves searching for the combination of values of the decision variables that realize the target values the objective functions. Different formulations and methods are described with precise mathematical details and illustrative numerical examples, covering mathematical programming, evolutionary algorithms, multi-objective optimization (MOO) and optimization under uncertainty, including robust optimization (RO).

Applications of the assessment and optimization methods to real-world cases are also given, concerning for example the reliability of renewable energy systems. From this point of view, the book bridges the gap between theoretical development and engineering practice.

Acknowledgments

Live long and prosper, RAMS and system reliability! The authors would express the deepest appreciations to the great scholars along the line of honors and achievements for their inspirations and role modeling.

Many thanks to the postgraduate students in Tsinghua: Tianli Men, Hanxiao Zhang, Ruochong Liu, Chen Zhang and Chuanzhou Jia. Thanks for their priceless efforts in editing, depicting, and proofreading in various chapters.

The authors would like to specially thank the Wiley colleagues for their continuous and kindhearted monitoring and encouragement throughout the years.

At last, this work is supported in part by the National Natural Science Foundation of China under a key project grant No. 71731008 and the Beijing Natural Science Foundation grant No. L191022.

List of Abbreviations

ABC	artificial bee colony algorithm
ACO	ant colony optimization
AGAN	as-good-as-new
B&B	branch-and-bound
BBA	basic belief assignment
BDD	binary decision diagram
BFS	basic feasible solution
BSS	binary state system
BSSPS	binary-state series-parallel system
cdf	cumulative distribution function
CG	column generation
CLT	Central Limit Theorem
CTMC	continuous time Markov chain
CVaR	conditional value-at-risk
DC	direct current
DE	deterministic equivalent
DE	differential evolution
DG	distributed generation
DM	decision maker
DP	dynamic programming
DTMC	discrete time Markov chain
EA	evolutionary algorithm
ENS	energy not supplied
EENS	expected energy not supplied
EV	electrical vehicles
FV	finite-volume
GA	genetic algorithm
GD	generational distance
HCTMC	homogeneous CTMC
HPIS	high-pressure injection system
HUGF	hybrid UGF
HV	hyper-volume

ICTMC	inhomogeneous CTMC
ILP	integer linear programming
IP	integer programming
LO	Linear optimization
LP	linear programming
LPM	LP master problem
MCMC	Markov Chain Monte Carlo
MCS	Monte Carlo simulation
MCS-OPF	Monte Carlo simulation – optimal power flow
MCV	minimal cut vector
MDD	multi-valued decision diagram
MH	Metropolis-Hastings
MIP	mixed integer programming
MOO	multi-objective optimization
MP	mathematical programming
MPV	minimal path vector
MRC	Markov renewal chain
MS	Main supply power spot
MSCS	multi-state coherent system
MSM	multi-state model
MSMS	multi-state monotone system
MSS	multi-state system
MTBF	mean time between failures
MTBR	mean time between repairs
MTTF	mean time to failure
NLP	non-linear programming
NPGA	niched Pareto GA
NPP	nuclear power plant
NSGA-II	fast non-dominated sorting genetic algorithm
OPF	optimal power flow
pdf	probability density function
PDMP	piecewise-deterministic Markov process
pmf	probability mass function
P-o-F	Physics-of-Failure
PSO	particle swarm optimization
PV	solar photovoltaic
RAM	reliability, availability, and maintainability
RAMS	RAM and Safety criteria
RAMS+C	RAMS and Cost
RAP	redundancy allocation problem
RC	robust counterpart
RESTART	Repetitive Simulation Trials After Reaching Thresholds
RFN	random-fuzzy number

RLPM	restricted LPM
RO	robust optimization
SMP	semi-Markov process
SODE	single-object DE
SOEA	single-objective EA
SOGA	single-objective GA
SOO	single-objective optimization
SOPSO	single-objective PSO
SP	stochastic programming
SPEA	strength Pareto evolutionary algorithm
SPEA 2	improved strength Pareto evolutionary algorithm
SSO	social spider optimization
ST	storage device
TDMSM	time-dependent MSM
TIMSM	time-independent MSM
TS	Tabu search
UGF	universal generating function
VEGA	vector-evaluated GA
W	wind turbine

Notations: Part I

t	time point
$n_f(t)$	number of failed items
$n_s(t)$	number of the survived items
n_0	sample size
T	random variable of the failure time
$F(t)$	cdf of failure time
$f(t)$	pdf of failure time
$R(t)$	reliability at time t
$h(t)$	hazard function at time t
$H(t)$	cumulative hazard function at time t
$\hat{Q}(t)$	estimate of the unreliability
$\hat{R}(t)$	estimate of the reliability
$D(t)$	component or system demand at time t
$G(t)$	performance function at time t
$MTTF$	mean time to failure
X	random variable

a	crack length
N	load cycle
Q	total volume of wear debris produced
$R_s(t)$	reliability of the system at time t
$(\cdot)$	unreliability function of the system
C	cost
x	decision variable
$\mathbf{g}(\boldsymbol{x})$	inequality constraints
$\boldsymbol{h}(\boldsymbol{x})$	equality constraints
$f(x)$	criterion function
$D=(V,A)$	directed graph
$d(\cdot)$	length of the shortest path

Notations: Part II

t	time point
S	state set
M	perfect state
$\boldsymbol{x}=(x_1,\ldots,x_n)$	component state vector
$\boldsymbol{X}=(X_1,\ldots,X_n)$	state of all components
$\phi(\cdot)$	structure function of the system
g_i	performance level of component i
λ_{kj}^i	transition rate of component i from state k to state j
$Q_{kj}^i(t)$	kernel of the SMP analogous to λ_{kj}^i of the CTMC
T_n^i	time of the n-th transition of component i
G_n^i	performance of component i at the n-th transition
$\theta_{jk}^i(t)$	probability that the process of component i starts from state j at time t
$A_\varphi^W(t)$	availability with a minimum on performance of total φ at time t
$u_i(z)$	universal generating function of component i
$p_{ij}=\Pr(X_i=j)$	probability of component i being at state j
$\boldsymbol{p}(t)$	state probability vector
$\lambda_{ij}(t)$	transition rate from state i to state j at time t in Markov process
$\boldsymbol{\Lambda}$	transition rate matrix
$\Pi(\cdot)$	possibility function
$N(\cdot)$	necessity function

$Bel(\cdot)$	belief function
$Pl(\cdot)$	plausibility function
$F^{-1}(\cdot)$	inverse function
$E(\cdot)$	expectation equation
$\otimes$	UGF composition operator
$\Pi(\cdot)$	possibility function
$N(\cdot)$	necessity function
$Bel(\cdot)$	belief function
$Pl(\cdot)$	plausibility function
$F^{-1}(\cdot)$	inverse function
$E(\cdot)$	expectation equation
$S(\cdot)$	system safety function
$Risk(\cdot)$	system risk function
$C(\cdot)$	cost function

Notation: Part III

r_i	reliability of subsystem i
$\boldsymbol{x}=(x_1,\ldots,x_n)^T$	decision variable vector
$\boldsymbol{c}=(c_1,\ldots,c_n)^T$	coefficients of the objective function
$\boldsymbol{b}=(b_1,\ldots,b_m)^T$	right-hand side values of the inequality constraints
$z=(z_1,z_2,\ldots,z_M)$	objective vector
$\boldsymbol{x}_l^*, l=1,2,\ldots,L$	set of optimal solutions
$w=(w_1,w_2,\ldots,w_M)$	weighting vector
$\boldsymbol{x}^*$	global optimal solution
$R(\cdot)$	system reliability function
$A(\cdot)$	system availability function
$M(\cdot)$	system maintainability function
	system safety function
$C(\cdot)$	cost function
$Risk(\cdot)$	system risk function
$\mathbf{R}^N$	N-dimensional solution space
f_i	i-th objective functions
g_j	j-th equality constraints
h_k	k-th inequality constraints

ω	random event
$\xi = \left(\mathbf{q}(\omega)^T, \mathbf{h}(\omega)^\mathrm{T}, T(\omega)^T\right)$	second-stage problem parameters
$\boldsymbol{W}$	recourse matrix
$\boldsymbol{y}(\omega)$	second-stage or corrective actions
$\mathcal{Q}(\boldsymbol{x})$	expected recourse function
$\mathcal{U}$	uncertainty set
$\boldsymbol{u}$	uncertainty parameters
ζ	perturbation vector
$\mathcal{Z}$	perturbation set
$\boldsymbol{x}_u^*$	optimal solution under the uncertainty parameter $\mathbf{u}$

Part I

The Fundamentals

1

Reliability Assessment

Reliability is a critical attribute for the modern technological components and systems. Uncertainty exists on the failure occurrence of a component or system, and proper mathematical methods are developed and applied to quantify such uncertainty. The ultimate goal of reliability engineering is to quantitatively assess the probability of failure of the target component or system [1]. In general, reliability assessment can be carried out by both parametric or nonparametric techniques. This chapter offers a basic introduction to the related definitions, models and computation methods for reliability assessments.

1.1 Definitions of Reliability

According to the standard ISO 8402, reliability is the ability of an item to perform a required function, under given environmental and operational conditions and for a stated period of time without failure. The term "item" refers to either a component or a system. Under different circumstances, the definition of reliability can be interpreted in two different ways:

1.1.1 Probability of Survival

Reliability of an item can be defined as the complement to its probability of failure, which can be estimated statistically on the basis of the number of failed items in a sample. Suppose that the sample size of the item being tested or monitored is n_0. All items in the sample are identical, and subjected to the same environmental and operational conditions. The number of failed items is n_f and the number of the survived ones is n_s, which satisfies

$$n_f + n_s = n_0 \tag{1.1}$$

Reliability Analysis, Safety Assessment and Optimization: Methods and Applications in Energy Systems and Other Applications, First Edition. Enrico Zio and Yan-Fu Li.

The percentage of the failed items in the tested sample is taken as an estimate of the unreliability, $\hat{Q}$,

$$\hat{Q} = \frac{n_f}{n_0} \tag{1.2}$$

Complementarily, the estimate of the reliability, $\hat{R}$, of the item is given by the percentage of survived components in the sample:

$$\hat{R} = \frac{n_s}{n_0} = 1 - \hat{Q} \tag{1.3}$$

Example 1.1

A valve fabrication plant has an average output of 2,000 parts per day. Five hundred valves are tested during a reliability test. The reliability test is held monthly. During the past three years, 3,000 valves have failed during the reliability test. What is the reliability of the valve produced in this plant according to the test conducted?

Solution

The total number of valves tested in the past three years is

$$n_0 = 500 \times 12 \times 3 = 18000$$

The number of failed components is

$$n_f = 3000$$

According to Equation 1.3, an estimate of the valve reliability is

$$\hat{R} = \frac{n_s}{n_0} = \frac{n_0 - n_f}{n_0} = \frac{18000 - 3000}{18000} \approx 0.833$$

1.1.2 Probability of Time to Failure

Let random variable T denote the time to failure. Then, the reliability function at time t can be expressed as the probability that the component does not fail at time t, that is,

$$R(t) = P(T > t) \tag{1.4}$$

Denote the cumulative distribution function (cdf) of T as $F(t)$. The relationship between the cdf and the reliability is

$$R(t) = 1 - F(t) \tag{1.5}$$

Further, denote the probability density function (pdf) of failure time T as $f(t)$. Then, equation (1.5) can be rewritten as

$$R(t)=1-\int_0^t f(\xi)d\xi \tag{1.6}$$

Example 1.2

The failure time of a valve follows the exponential distribution with parameter $\lambda=0.025$ (in arbitrary units of time^{-1}). The value is new and functioning at time $t=0$. Calculate the reliability of the valve at time $t=30$ (in arbitrary units of time).

Solution

The pdf of the failure time of the valve is

$$f(t)=\lambda e^{-\lambda t}=0.025e^{-0.025t}, t\geq 0$$

The reliability function of the valve is given by

$$R(t)=1-\int_0^t 0.025e^{-0.025\xi}d\xi$$

At time $t=30$, the value of the reliability is

$$R(30)=1-\int_0^{30} 0.025e^{-0.025\xi}d\xi\approx 0.472$$

In all generality, the expected value or mean of the time to failure T is called the *mean time to failure* (MTTF), which is defined as

$$MTTF=E[T]=\int_0^\infty tf(t)dt \tag{1.7}$$

It is equivalent to

$$MTTF=\int_0^\infty R(t)dt \tag{1.8}$$

Another related concept is the *mean time between failures* (MTBF). MTBF is the average working time between two consecutive failures. The difference between MTBF and MTTF is that the former is used only in reference to a repairable item, while the latter is used for non-repairable items. However, MTBF is commonly used for both repairable and non-repairable items in practice.

The failure rate function or hazard rate function, denoted by $h(t)$, is defined as the conditional probability of failure in the time interval $[t, t+\Delta t]$ given that it has been working properly up to time t, which is given by

$$h(t) = \lim_{\Delta t \to 0} P(T \le t + \Delta t \mid T > t) = \frac{f(t)}{R(t)} \tag{1.9}$$

Furthermore, the cumulative failure rate function, or cumulative hazard function, denoted by $H(t)$, is given by

$$H(t) = \int_0^t h(t)dt \tag{1.10}$$

1.2 Component Reliability Modeling

As mentioned in the previous section, in reliability engineering, the time to failure of an item is a random variable. In this section, we briefly introduce several commonly used discrete and continuous distributions for component reliability modeling.

1.2.1 Discrete Probability Distributions

If random variable X can take only a finite number k of different values $x_1, x_2, \ldots, x_k$ or an infinite sequence of different values $x_1, x_2, \ldots$, the random variable X has a discrete probability distribution. The probability mass function (pmf) of X is defined as the function f such that for every real number x,

$$f(x) = P(X = x) \tag{1.11}$$

If x is not one of the possible values of X, then $f(x) = 0$. If the sequence $x_1, x_2, \ldots$ includes all the possible values of X, then $\sum_i f(x_i) = 1$. The cdf is given by

$$F(x_i) = P(X \le x_i) \tag{1.12}$$

1.2.1.1 Binomial Distribution

Consider a machine that produces a defective item with probability p $(0 < p < 1)$ and produces a non-defective item with probability $1 - p$. Assume the events of defects in different items are mutually independent. Suppose the experiment consists of examining a sample of n of these items. Let X denote the number of defective items in the sample. Then, the random variable X follows a binomial distribution with parameters n and p and has the discrete distribution represented by the pmf in (1.14), shown in Figure 1.1. The random variable with this distribution is said to be a binomial random variable, with parameters n and p,

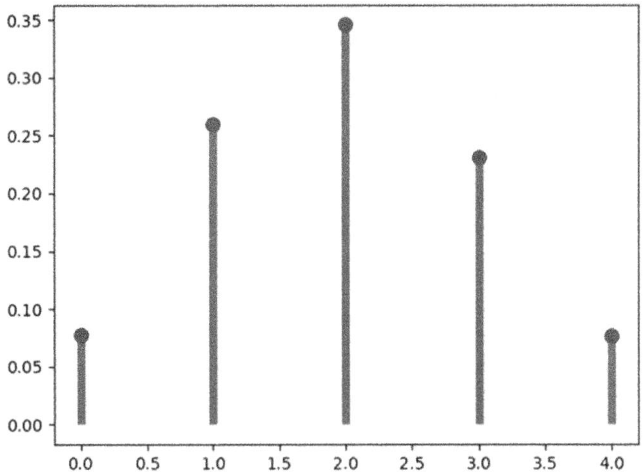

Figure 1.1 The pmf of the binomial distribution with $n=5,\ p=0.4$.

$$f(x)=\begin{cases}\binom{n}{x}p^x(1-p)^{n-x}, & for\, x=0,1,\ldots,n,\\ 0, & otherwise.\end{cases} \tag{1.13}$$

The pmf of the binomial distribution is

$$F(x)=\sum_{i=0}^{x}\binom{n}{i}p^i(1-p)^{n-i} \tag{1.14}$$

For a binomial distribution, the mean, μ, is given by

$$\mu=np \tag{1.15}$$

and the variance, σ^2, is given by

$$\sigma^2=np(1-p) \tag{1.16}$$

1.2.1.2 Poisson Distribution

Poisson distribution is widely used in quality and reliability engineering. A random variable X has the Poisson distribution with parameter $\lambda,\lambda>0$, the pmf (shown in Figure 1.2) of X is as follows:

$$f(x)=\begin{cases}\dfrac{e^{-\lambda}\lambda^x}{x!}, & for\, x=0,1,\ldots,\\ 0, & otherwise.\end{cases} \tag{1.17}$$

The mean and variance of the Poisson distribution are

$$\mu=\sigma^2=\lambda \tag{1.18}$$

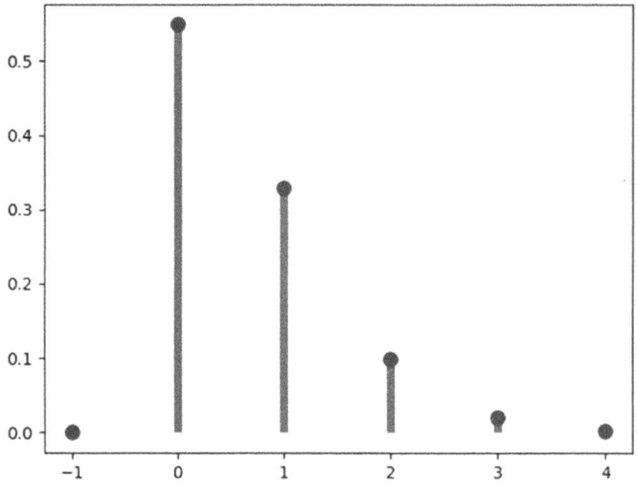

Figure 1.2 The pmf of the Poisson distribution with $\lambda = 0.6$.

1.2.2 Continuous Probability Distributions

We say that a random variable X has a continuous distribution or that X is a continuous random variable if there exists a nonnegative function f, defined on the real line, such that for every interval of real numbers (bounded or unbounded), the probability that X takes a value in an interval $[a,b]$ is the integral of f over that interval, that is,

$$P(a \leq X \leq b) = \int_a^b f(x)dx. \tag{1.19}$$

If X has a continuous distribution, the function f will be the probability density function (pdf) of X. The pdf must satisfy the following requirements:

$$f(x) \geq 0, \text{ for all } x. \tag{1.20}$$

The cdf of a continuous distribution is given by

$$\int_{-\infty}^{\infty} f(x)dx = 1. \tag{1.21}$$

The mean, μ, and variance, σ^2, of the continuous random variable are calculated by

$$\mu = \int_{-\infty}^{\infty} x f(x) dx \tag{1.22}$$

$$\sigma^2 = \int_{-\infty}^{\infty} (x-\mu)^2 f(x) dx.$$

1.2.2.1 Exponential Distribution

A random variable T follows the exponential distribution if and only if the pdf (shown in Figure 1.3) of T is

$$f(t) = \lambda e^{-\lambda t}, t \geq 0, \tag{1.23}$$

where $\lambda > 0$ is the parameter of the distribution. The cdf of the exponential distribution is

$$F(t) = 1 - e^{-\lambda t}, t \geq 0. \tag{1.24}$$

If T denotes the failure time of an item with exponential distribution, the reliability function will be

$$R(t) = e^{-\lambda t}, t \geq 0. \tag{1.25}$$

The hazard rate function is

$$h(t) = \lambda. \tag{1.26}$$

The mean, μ, and variance, σ^2 are

$$\mu = \frac{1}{\lambda} \tag{1.27}$$

$$\sigma^2 = \frac{1}{\lambda^2}.$$

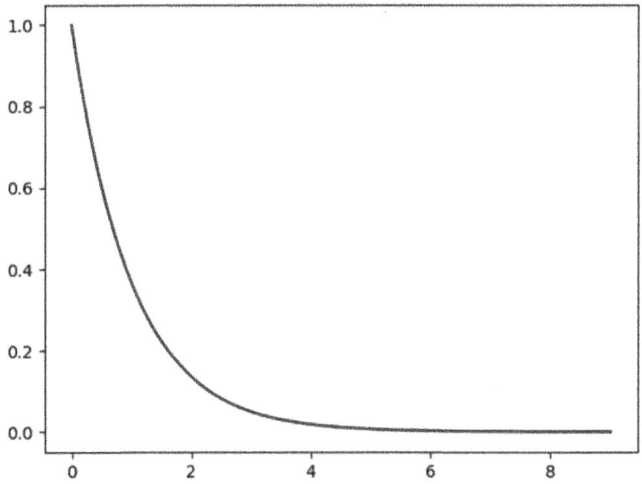

Figure 1.3 The pdf of the exponential distribution with $\lambda = 1$.

1.2.2.2 Weibull Distribution

A random variable T follows the Weibull distribution if and only if the pdf (shown in Figure 1.4) of T is

$$f(t) = \frac{\beta t^{\beta-1}}{\eta^{\beta}} e^{-\left(\frac{t}{\eta}\right)^{\beta}}, t \geq 0, \tag{1.28}$$

where $\beta > 0$ is the shape parameter and $\eta > 0$ is the scale parameter of the distribution. The cdf of the Weibull distribution is

$$F(t) = 1 - e^{-\left(\frac{t}{\eta}\right)^{\beta}}, t \geq 0. \tag{1.29}$$

If T denotes the time to failure of an item with Weibull distribution, the reliability function will be

$$R(t) = e^{-\left(\frac{t}{\eta}\right)^{\beta}}, t \geq 0. \tag{1.30}$$

The hazard rate function is

$$h(t) = \frac{\beta}{\eta}\left(\frac{t}{\eta}\right)^{\beta-1}, t \geq 0. \tag{1.31}$$

The mean, μ, and variance, σ^2, are

$$\mu = \eta \Gamma\left(\frac{1+\beta}{\beta}\right), \tag{1.32}$$

$$\sigma^2 = \eta^2 \left[\Gamma\left(\frac{2+\beta}{\beta}\right) - \left(\Gamma\left(\frac{1+\beta}{\beta}\right)\right)^2\right].$$

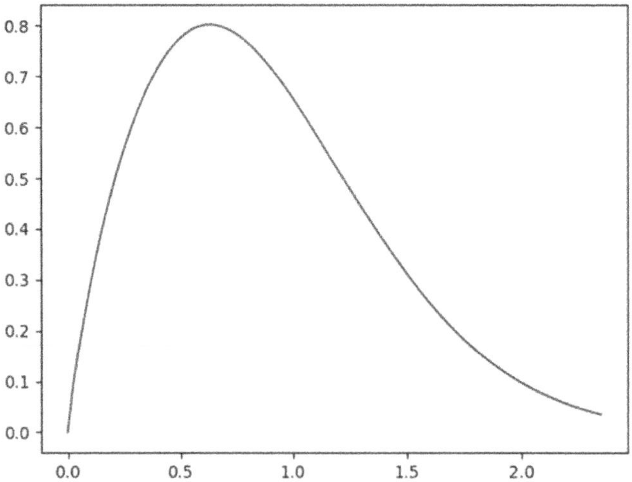

Figure 1.4 The pdf of the Weibull distribution with $\beta = 1.79$, $\eta = 1$.

1.2.2.3 Gamma Distribution

A random variable T follows the gamma distribution if and only if the pdf (shown in Figure 1.5) of T is

$$f(t) = \frac{\lambda^{\beta}}{\Gamma(\beta)} t^{\beta-1} e^{-\lambda t}, t \geq 0, \tag{1.33}$$

where $\beta > 0$ is the shape parameter and $\eta > 0$ is the scale parameter of the distribution. The cdf of the gamma distribution is

$$F(t) = \frac{\lambda^{\beta}}{\Gamma(\beta)} \int_0^t x^{\beta-1} e^{-\lambda x} dx, t \geq 0. \tag{1.34}$$

If T denotes the failure time of an item with gamma distribution, the reliability function will be

$$R(t) = \frac{\lambda^{\beta}}{\Gamma(\beta)} \int_t^{\infty} x^{\beta-1} e^{-\lambda x} dx, t \geq 0. \tag{1.35}$$

The hazard rate function is

$$h(t) = \frac{t^{\beta-1} e^{-\lambda t}}{\int_t^{\infty} x^{\beta-1} e^{-\lambda x} dx}, t \geq 0. \tag{1.36}$$

The mean, μ, and variance, σ^2, are

$$\mu = \frac{\beta}{\lambda} \qquad \sigma^2 = \frac{\beta}{\lambda^2}. \tag{1.37}$$

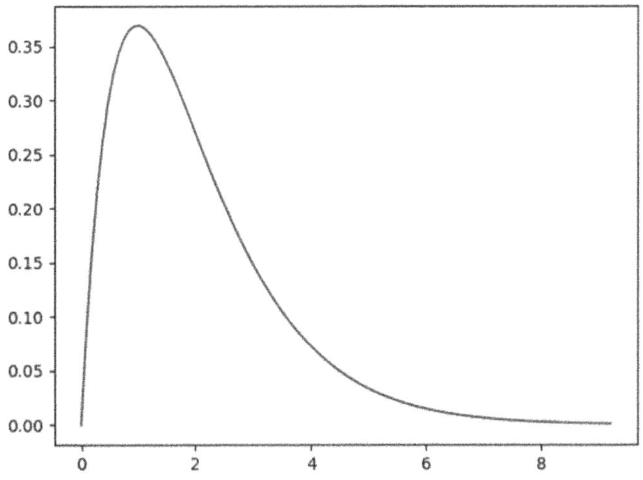

Figure 1.5 The pdf of the gamma distribution with $\beta = 1.99$, $\lambda = 1$.

1.2.2.4 Lognormal Distribution

A random variable T follows the lognormal distribution if and only if the pdf (shown in Figure 1.6) of T is

$$f(t) = \frac{1}{\sigma t\sqrt{2\pi}} \exp\left[-\frac{1}{2\sigma^2}(\ln t - \mu)^2\right], t > 0, \tag{1.38}$$

where $\sigma > 0$ is the shape parameter and $\mu > 0$ is the scale parameter of the distribution. Note that the lognormal variable is developed from the normal distribution. The random variable $X = \ln T$ is a normal random variable with parameters μ and σ. The cdf of the lognormal distribution is

$$F(t) = \Phi\left(\frac{\ln t - \mu}{\sigma}\right), t > 0, \tag{1.39}$$

where $\Phi(x)$ is the cdf of a standard normal random variable. If T denotes the failure time of an item with lognormal distribution, the reliability function of T will be

$$R(t) = 1 - \Phi\left(\frac{\ln t - \mu}{\sigma}\right), t > 0. \tag{1.40}$$

The hazard rate function is

$$h(t) = \frac{f(t)}{1 - \Phi\left(\frac{\ln t - \mu}{\sigma}\right)}, t > 0. \tag{1.41}$$

The mean, μ, and variance, σ^2, are

$$\mu = e^{\mu + \sigma^2/2}, \quad \sigma^2 = e^{2\mu + \sigma^2}\left(e^{\sigma^2} - 1\right). \tag{1.42}$$

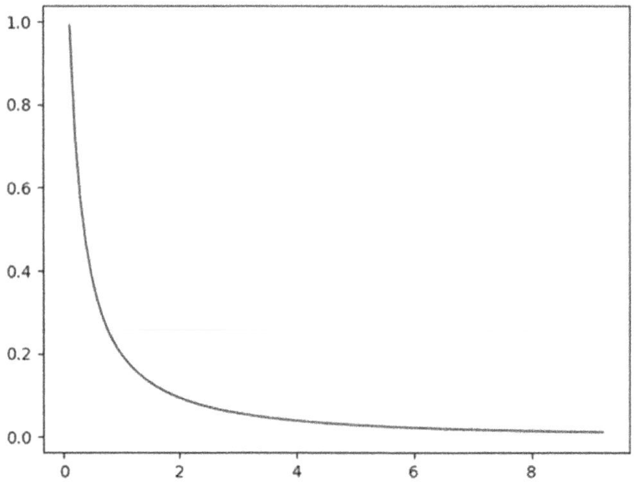

Figure 1.6 The pdf of the lognormal distribution with $\mu = 0$, $\sigma = 0.954$.

Example 1.3

The random variable of the time to failure of an item, T, follows the following pdf:

$$f(t) = \begin{cases} \frac{1}{6000}, & 0 \le t \le 6000, \\ 0, & otherwise. \end{cases}$$

where t is in days and $t \ge 0$.

a) What is the probability of failure of the item in the first 100 days?
b) Find the MTTF of the item.

Solution

a) The cdf of the random variable is

$$F(t) = \begin{cases} \frac{t}{6000}, & 0 \le t \le 6000, \\ 0, & otherwise. \end{cases}$$

The probability of failure in the first 100 days is

$$P(T \le 100) = F(100) = \frac{100}{6000} \approx 0.017.$$

b) The MTTF of the item is

$$\text{MTTF} = \text{E}[T] = \int_0^{6000} \frac{6000 - t}{6000} dt = 3000 \text{ days}.$$

1.2.3 Physics-of-Failure Equations

Different from the traditional reliability assessment approach, the Physics-of-Failure (P-o-F) represents an approach to reliability assessment based on modeling and simulation of the physical processes leading to the occurrence of failures in an item [2]. The P-o-F approach begins within the first stages of the design of the item. A model is constructed based on the customer's requirements, service environment, and stress analysis [1]. Once the models are established, a reliability assessment can be conducted on the item.

1.2.3.1 Paris' Law for Crack Propagation

Paris' law is a crack growth equation that gives the rate of growth of a fatigue crack [3]. The stress intensity factor K characterizes the load around a crack tip and the rate of crack growth is experimentally shown to be a function of the range of the stress intensity ΔK experienced in a loading cycle (shown in Figure 1.7). The Paris' equation describing this is

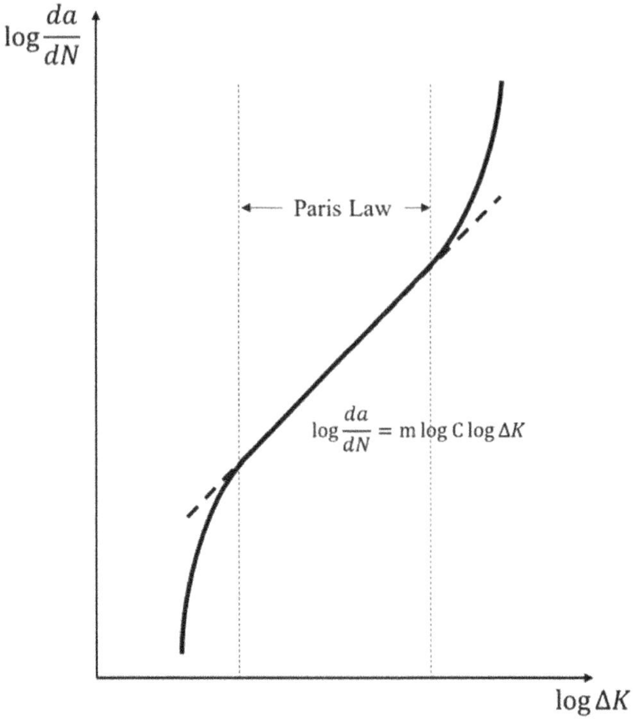

Figure 1.7 Illustration of Paris Law.

$$\frac{da}{dN} = C(\Delta K)^m, \tag{1.43}$$

where a is the crack length and $\frac{da}{dN}$ is the fatigue crack growth for a load cycle N. The material coefficients C and m are obtained experimentally and their values depend on environment, temperature, and stress ratio. The stress intensity factor range has been found to correlate with the rate of crack growth in a variety of different conditions, which is the difference between the maximum and minimum stress intensity factors in a load cycle, defined as

$$\Delta K = K_{max} - K_{min}. \tag{1.44}$$

1.2.3.2 Archard's Law for Wear

The Archard's wear equation is a simple model used to describe sliding wear, which is based on the theory of asperity contact [4]. The volume of the removed debris due to wear is proportional to the work done by friction forces. The Archard's wear equation is given by

$$Q = \frac{KWL}{H}, \tag{1.45}$$

where Q is the total volume of the wear debris produced, K is a dimensionless constant, W is the total normal load, L is the sliding distance, and H is the hardness of the softest contacting surfaces. It is noted that WL is proportional to the friction forces. K is obtained from experimental results and it depends on several parameters, among which are surface quality, chemical affinity between the material of two surfaces, surface hardness process, etc.

1.3 System Reliability Modeling

The methods to model and estimate the reliability of a single component were introduced in Section 1.2. Compared with the single component case, the system reliability modeling and assessment is more complicated. The term 'system' is used to indicate a collection of components working together to perform a specific function. The reliability of a system depends not only on the reliability of each component but also on the structure of the system, the interdependence of its components, and the role of each component within the system, etc. To compute the reliability of the system, it is essential to construct the model of the system, representing the above characteristics.

The conventional approaches typically assume that the components and the system have two states: perfect working and complete failure [5]. Below, we introduce the reliability models of a binary state system with specific structures. Details about the multi-state system can be found in Chapter 3.

1.3.1 Series System

In a series system, all components must operate successfully for the system to function or operate successfully. It implies that the failure of any component will cause the entire system to fail. The reliability block diagram of a series system is shown in Figure 1.8.

Let R_i be the reliability of the ith component, $i = 1,2,\ldots,n,$, and R_s be the reliability of the system. Let x_i be the event that the ith component is operational and let x be the event that indicates system is operational. The reliability of the series system can be calculated by

$$R_s = P(x) = P(x_1, x_2, \ldots, x_n). \tag{1.46}$$

Assume all the components in the series system are independent; if so, the reliability of the system can be expressed as

$$R_s = \prod_{i=1}^{n} R_i. \tag{1.47}$$

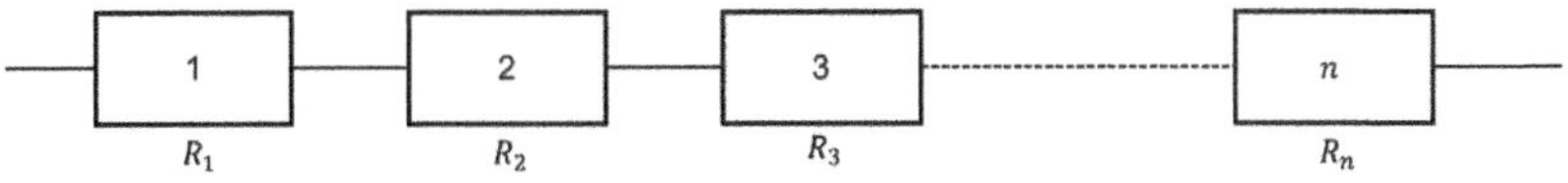

Figure 1.8 Reliability block diagram of a series system.

Considering that the component reliability is a number between 0 and 1, we have the following relationship

$$R_s <= min\{R_1, R_2, \ldots, R_n\}. \tag{1.48}$$

1.3.2 Parallel System

In a parallel system, the system functions or operates successfully when at least one component function is working. It implies that the failure of all components will cause the entire system to fail. The reliability block diagram of a parallel system is shown in Figure 1.9.

Denote F_s as the probability of failure of the system. Denote F_i as the probability of failure of component i. The system reliability can be expressed as

$$R_s = 1 - F_s = 1 - \prod_{i=1}^{n} F_i = 1 - \prod_{i=1}^{n} [1 - R_i]. \tag{1.49}$$

It follows that

$$R_s \geq max\{R_1, R_2, \ldots, R_n\}. \tag{1.50}$$

1.3.3 Series-parallel System

A series-parallel system consists of m subsystems that are connected in series, with n_i units connected in parallel in each subsystem, $i = 1, \ldots, m$. The reliability block diagram of a series-parallel system is shown in Figure 1.10.

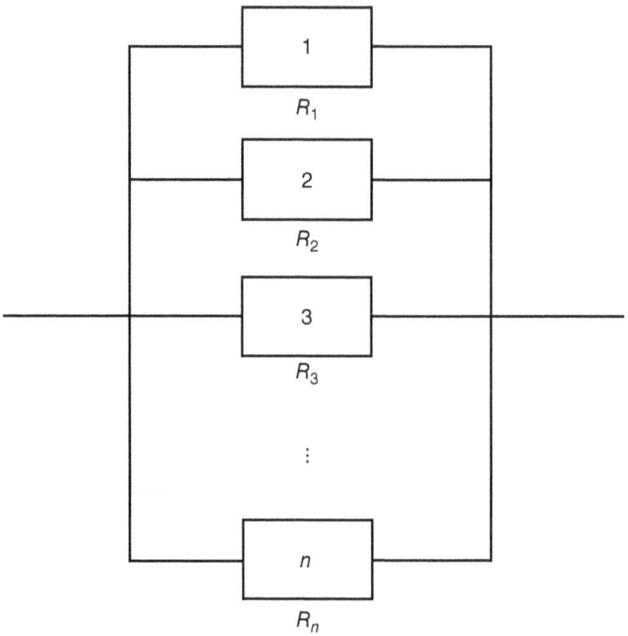

Figure 1.9 Reliability block diagram of a parallel system.

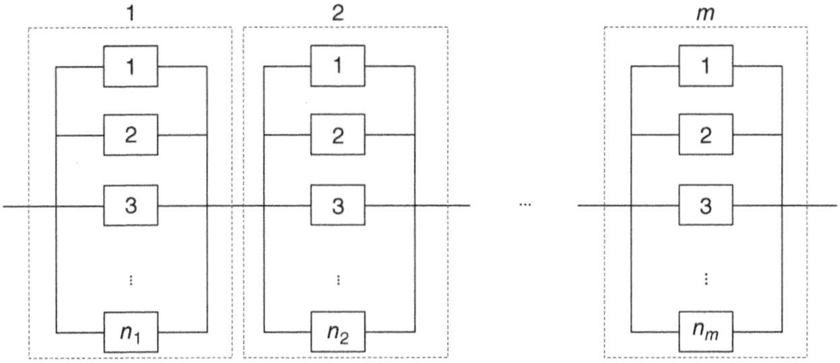

Figure 1.10 Reliability block diagram of a series-parallel system.

Denote R_{ij} as the reliability of component j in subsystem i, $1 \le i \le m, 1 \le j \le n_i$. Let R_i be the reliability of the subsystem i, $1 \le i \le m$. First, the reliability of each subsystem is derived as for the parallel system, that is,

$$R_i = 1 - \prod_{j=1}^{n_i} \left(1 - R_{ij}\right), i = 1, 2, \ldots, m. \tag{1.51}$$

The reliability of the series-parallel system is, then,

$$R_s = \prod_{i=1}^{m} R_i = \prod_{i=1}^{m} \left(1 - \prod_{j=1}^{n_i} \left(1 - R_{ij}\right)\right). \tag{1.52}$$

1.3.4 K-out-of-n System

For a system composed of n components, the system is operational if and only if at least k of the n components are operational. We call this type of system as k-out-of-n: G system, where G is short for Good. For a system composed of n components, the system fails if and only if at least k of the n components are failed. We call this type of system a k-out-of-n: F system. According to the definition, the series system is a 1-out-of-n: F system, where F is short for Failed. The parallel system is a 1-out-of-n: G system. We will mainly present the reliability of the k-out-of-n: G system here.

Assume that the n components are identical and independent. Denote R as the reliability of each component, F as the unreliability of each component, $F = 1 - R$. Let P_i be the probability so that exactly i components are functional. In a k-out-of-n: G system, the number of functional components follows the binomial distribution with parameter n and R. The probability that exactly i components are functional, P_i, is

$$P_i = \binom{n}{i} R^i F^{n-i}, i = 0, 1, 2, \ldots, n \tag{1.53}$$

The reliability of the system is the probability that the number of functional components is greater than or equal to k. Thus, the system reliability, R_s, is calculated by

$$R_s = \sum_{i=k}^{n} P_i = \sum_{i=k}^{n} \binom{n}{i} R^i F^{n-i}. \tag{1.54}$$

If the components are not identical, the system reliability should be calculated by enumerating all combinations of working components.

1.3.5 Network System

There are systems that can be represented by network diagrams, for example, gas networks, telecommunications networks, and power networks. A network system consists of a set of nodes and links. All the nodes and links have a probability of failure.

1.4 System Reliability Assessment Methods

There are many reliability assessment approaches developed to compute the reliability of complex systems, e.g. networks. Path-set and cut-set methods, decomposition and factorization methods, and binary decision diagram (BDD) are four commonly used methods, and we will introduce them in this section.

1.4.1 Path-set and Cut-set Method

A path set P is a set of components, which by functioning ensures that the system is functioning. A path set is said to be minimal if it cannot be reduced without losing its status as a path set. A cut set K is a set of components, which by failing causes the system to fail. A cut set is said to be minimal if it cannot be reduced without losing its status as a cut set. We refer to these minimal sets as minimal path and cut sets or vectors (MPSs, MPVs and MCSs, MCVs).

Consider the minimal path sets of the system, P_1, P_2,..., P_p, and the minimal cut sets of the system, K_1, K_2,..., K_k. The reliability of the system is given by the union of all minimal path sets. The unreliability is given by the probability that at least one minimal cut set occurs.

Example 1.4

Consider a bridge structure with five edges, E_1,..., E_5, as shown in Figure 1.11:

a) Find the minimal path sets and the minimal cut sets of the system.
b) Calculate the reliability of the system if the reliability of each component is R.

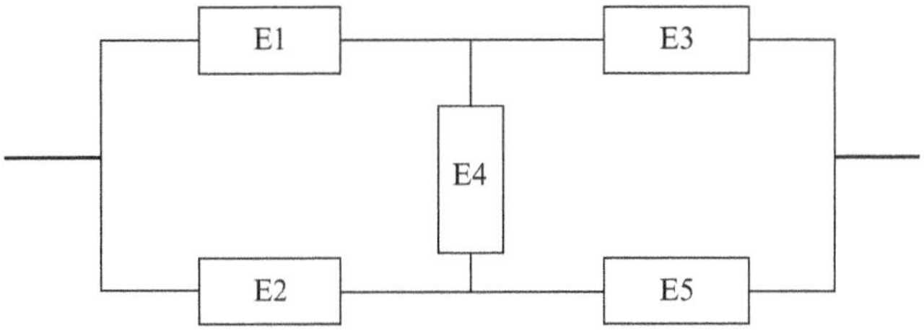

Figure 1.11 Bridge system.

Solution

a) The minimal path sets are

$$P_1 = \{1,3\}, P_2 = \{2,5\}, P_3 = \{1,4,5\}, P_4 = \{2,3,4\}.$$

The minimal cut sets are

$$K_1 = \{1,2\},\ K_2 = \{3,5\},\ K_3 = \{1,4,5\},\ K_4 = \{2,3,4\}.$$

b) The reliability of the system is calculated by the union of the path sets:

$$= 2R^2 + 2R^3 - 5R^4 + 2R^5.$$

1.4.2 Decomposition and Factorization

The decomposition method begins by selecting a critical component, denoted by x, which is an important component of the complex system structure. The reliability of the system can be calculated by the conditional probability:

$$R_s = P(\text{system functional}|x)R(x) + P(\text{system functional} \mid \bar{x})(1 - R(x)). \quad (1.55)$$

The factorization method is developed based on the decomposition method, which is used in a network system. Denote e as a critical edge in the network G. The reliability of the network is

$$R_s = P(G\ \text{functional}|e)R_e + P(G\ \text{functional} \mid \bar{e})(1 - R_e). \quad (1.56)$$

1.4.3 Binary Decision Diagram

Binary decision diagram (BDD) is used to represent a Boolean function. A Boolean function can be represented as a rooted, directed, acyclic graph, which consists of several nodes and two terminal nodes. The two terminal nodes are labeled 0 (FALSE) and 1 (TRUE). Each node u is labeled by a Boolean variable x_i and has two child nodes called low child and high child. The edge from a node to a child represents an assignment of the value FALSE (or TRUE, respectively) to variable x_i. The advantage of BDD in reliability assessment is that its accuracy and efficiency are high [6]. The algorithm to compute the probability of a gate from a BDD is based on the Shannon Decomposition, which is defined by recursive equations.

Example 1.5

Calculate the reliability of the bridge system in Figure 1.11, if the reliability of each component is R.

Solution

The block decision diagram of the bridge system is shown in Figure 1.12.
The reliability of the system is

$$R_s = R^3 + R^3(1-R) + R^4(1-R) + 2R^3(1-R)^2 + R^2(1-R) + R^3(1-R)^2$$
$$+R^3(1-R)^2 + R^2(1-R)^3 = 2R^2 + 2R^3 - 5R^4 + 2R^5.$$

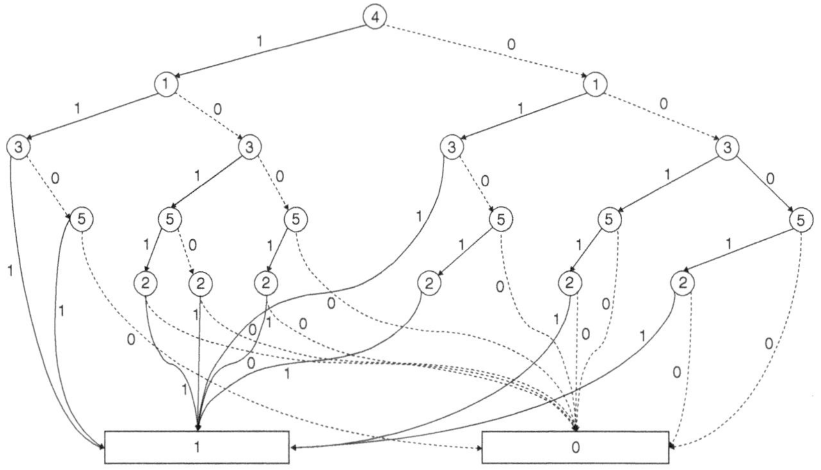

Figure 1.12 Block decision diagram of the bridge system.

1.5 Exercises

1 Consider an electrical generating system with two engines, E_1, E_2, and three generators, G_1, G_2, G_3, each one with rate equal to 30 kVA. The system fails when the generators fail to supply at least 60 kVA. The structure of the system is shown in Figure 1.13.
   a. Find the minimal cut sets of the system.
   b. Estimate the unreliability of the system for one-month operation, given that the failure rate for each engine is $5\times10^{-6}h^{-1}$ and that for each generator is $10^{-5}h^{-1}$.

2 Consider the reliability of the following system consisting of five components in Figure 1.14. All the components are identical and independent from each other. The reliability of components i is R_i. Let R_s be the reliability of the system. Give the reliability formulation of the system.

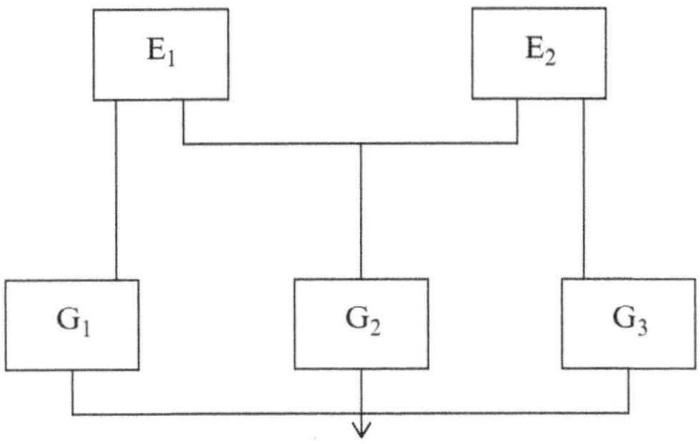

Figure 1.13 Electrical generating system.

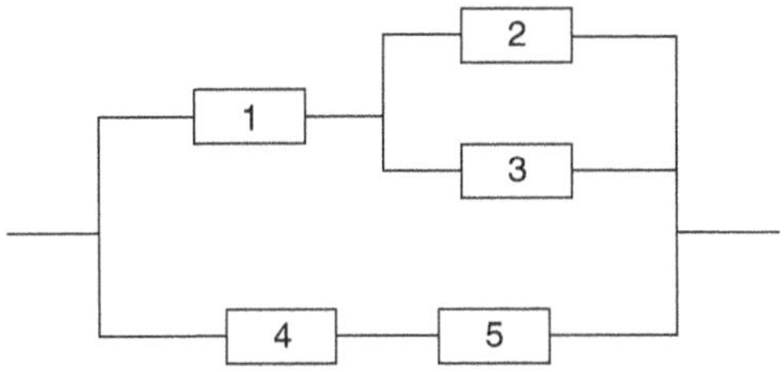

Figure 1.14 Reliability block diagram of the system.

3 The system has $N = 4$ components. Each component has three states: ($M \in \{0,1,2\}$). Let x_i denote the state of component i: then, we have the probability $P(x_i \geq 1) = 0.7$, $P(x_i = 2) = 0.5$, for $i = 1,2,3,4$. Give the following system structure function,

a. $\phi(x) = \min(x_1,(x_2 + x_3),x_4)$.

b. Find all minimal path and cut vectors (MPVs and MCVs) of the system.

c. Calculate system reliability $R = \Pr(\phi(x) \geq 1)$.

4 The power grid structure is shown in Figure 1.15 below. There are three substations: A is the power supplier that generates electric power to be transmitted to the substations B and C, which are the power consumers. Assume that the substations are always working but the power transmission lines may fail. The overall power grid works only if all the following conditions are satisfied:

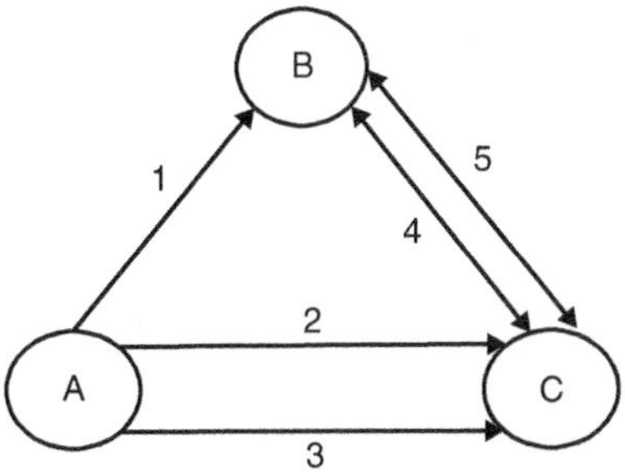

Figure 1.15 Diagram of the power grid structure.

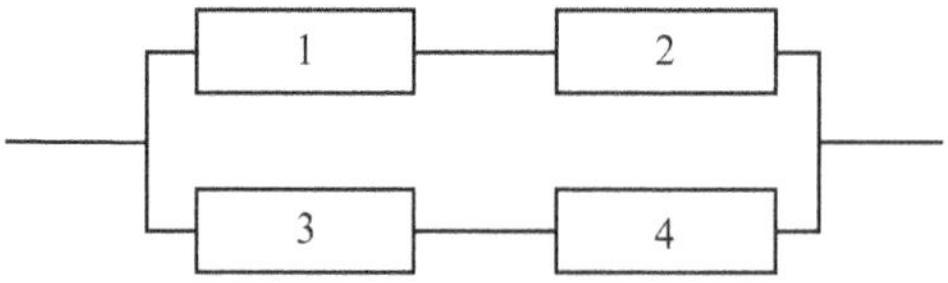

Figure 1.16 Reliability block diagram of the system.

i. Both substations B and C have power input.

ii. At least two outgoing transmission lines of A are working.

Then

a. Build a BDD for the power grid system.
b. Estimate the unreliability of the system for one-month operation by BDD, given that the failure rate for lines 1, 2, 3 is $\lambda_1 = 510^{-6}$ h^{-1} and for lines 4, 5 is $\lambda_2 = 10^{-5}$ h^{-1}.

5 Consider the series-parallel system in Figure 1.16. The components 1, 2, 3, and 4 are independent from each other and have exponential reliabilities with failure rates λ_1, λ_2, λ_3 and λ_4, respectively. Assuming that λ_1 and $\lambda_4 = \lambda_2$ /, calculate the system mean time to failure (MTTF) expression in terms of λ_2 and λ_3.

6 A manufacturer performs a test on a ceramic capacitor and finds that it experiences failures exponentially distributed in time, with rate $\lambda = 510^{-4}$ failures per hour. To retain operation performance of the ceramic capacitor, an instantaneous and imperfect maintenance activity is performed at an interval of 10^3 hours. The reliability after maintenance is 0.98. Calculate the average availability and the instantaneous availability at time 1.210^3 hours.

References

1 Zio, E. (2007). *An Introduction to the Basics of Reliability and Risk Analysis*, Vol. 13. World scientific.
2 Matic, Z. and Sruk, V. (2008, June). The physics-of-failure approach in reliability engineering. In *ITI 2008-30th International Conference on Information Technology Interfaces* (pp. 745–750). IEEE.
3 Paris, P. and Erdogan, F. (1963). A critical analysis of crack propagation laws.
4 Magnee, A. (1995). Generalized law of erosion: Application to various alloys and intermetallics. *Wear* 181: 500–510.
5 Barlow, R.E. and Proschan, F. (1975). *Statistical Theory of Reliability and Life Testing: Probability Models*. Florida State Univ Tallahassee.
6 Rauzy, A. (2008). Binary decision diagrams for reliability studies. In: *Handbook of Performability Engineering*, (ed. K.B. Misra) 381–396. London: Springer.

2

Optimization

Reliability optimization aims at maximizing system reliability and related metrics (e.g. weight and volume) while minimizing the cost required for the improvements of them. Reliability optimization has been an active research domain since the 1960s, with various formulations and solution schemes proposed. In general, the decision variables of the optimization problems are the parameters, which can be used for system reliability improvement. For instances, the parameters that define the system reliability allocation (e.g. component failure probability, failure rate), the parameters that describe system logic configuration (e.g. number of redundant components, component assignments), and those relevant to testing and maintenance activities (e.g. test intervals, maintenance periodicities).

In Section 2.1, four different types of reliability optimization problems are reviewed. The types are distinguished according to the nature of the decision variables. On the other hand, system reliability can be optimized through either single objective or multi-objective approach. The solution techniques to single-objective reliability optimization problems have been well documented in the surveys by Kuo and Prasad [1] and by Kuo and Wan [2]. We also reviewed the details of multi-objective reliability optimization problems in Chapter 10.

2.1 Optimization Problems

2.1.1 Component Reliability Enhancement

The objective of component reliability enhancement problems is to optimize system reliability via improving reliability metrics of individual components. In 1973, Kulshrestha and Gupta [3] first formulated one such problem to maximize the reliability of a series system, as follows

$$\max R = \prod_{i=1}^{N} r_i \tag{2.1a}$$

Reliability Analysis, Safety Assessment and Optimization: Methods and Applications in Energy Systems and Other Applications, First Edition. Enrico Zio and Yan-Fu Li.

$$\text{s.t.} \sum_{i=1}^{N} h_{ji}(r_i) \le b_j, \quad j = 1,2,\ldots,m \tag{2.1b}$$

where N is the number of subsystems, r_i is the reliability of subsystem i (i.e., component i, because each subsystem i is composed of only one component), $h_{ji}(r_i)$ is the j-th resource consumed at subsystem i, and b_j is the total amount of resource j available. This problem is also referred to as the reliability allocation problem, which is among the earliest attempts to system reliability optimization. Tillman et al. record the related publications during 1960s and 1970s in [4].

In literature, the objective of various research works are to optimize time-related reliability metrics, such as the mean time between failure (MTBF), the and mean time between repair (MTBR), and other lifetime distribution parameters of the components [5–6]. To achieve superior optimization results, component reliability enhancement is increasingly combined with other reliability improvement approaches, such as redundancy allocation. For example, in [7], the component reliability metrics, i.e. the component failure rates, repair rates, and component reliability, are regarded as the decision variables together with the number of redundancies in each subsystem. This type of problem is referred to as reliability-redundancy allocation problem [2].

2.1.2 Redundancy Allocation

The redundancy allocation, first formulated by Ghare and Taylor [8] in 1969 and Beraha and Misra [9] in 1974, is a well-established approach for reliability optimization. It aims to improve system reliability via installing additional redundant components into the system. A classical formulation of the redundancy allocation problem (RAP) is presented as follows, which aims to minimize the total system cost while keeping the system reliability R equal to or above a predefined acceptable level R_0.

$$\min C = \sum_{i=1}^{N} \sum_{j=1}^{v_i} c_{ij} y_{ij} \tag{2.2a}$$

$$\text{s.t.} R(y) = \prod_{i=1}^{N} \left(1 - \prod_{j=1}^{v_i} (1 - r_{ij})^{y_{ij}} \right) \ge R_0 \tag{2.2b}$$

$$u_{ij} \ge y_{ij} \ge l_{ij};\ y_{ij} \in \mathbb{N}_{\ge 0} \tag{2.2c}$$

The formulation is for a representative binary-state series-parallel system (BSSPS), where v_i is the number of component versions available to the i-th subsystem, r_{ij} is the reliability of the j-th component version in the i-th subsystem, $y = (y_{11}, \ldots, y_{1v_1}; \ldots; y_{N1}, \ldots, y_{Nv_N})$ is the decision vector, y_{ij} is the number of components of the j-th version in the i-th subsystem, and u_{ij} and l_{ij} are the upper and lower limits of the number of j-th component versions at the i-th subsystem, respectively.

An other classical formulation of RAP is to maximizes system reliability while keeping the cost below a certain budget.

RAP is an NP-hard [10] problem with non-linear and combinatorial nature. Most of the existing RAP works are based on a binary-state system (BSS) model. In literature, numerous methods have been proposed to solve it, including the exact methods [11,12] and the heuristic methods [13,14]. Comprehensive reviews on BSS RAP and its optimization solution methods can be found in Kuo and Prasad [1] and Kuo and Wan [2]. The MSS model has recently gained increasing popularity for system reliability assessment, because it realistically considers more than one intermediate state for the system and its elements between the two extremes of perfect functioning and complete failure. The MSS RAP was first investigated by Ushakov in 1987 [15] where the Universal Generating Function (UGF) approach [16] was used for reliability computation. Complex MSS RAP is typically solved by meta-heuristics, including genetic algorithm (GA) [17], tabu search (TS) [14], ant colony optimization (ACO) [18], particle swarm optimization (PSO) [19] and artificial bee colony (ABC) algorithm [20].

Similar to the case of component enhancement, redundancy allocation is also used in combination with other reliability improvement methods, such as maintenance and testing. In [21], joint redundancy and imperfect maintenance strategy optimization are considered. In [22], redundancy and number of maintenance teams are optimized together. In [23], redundancy and the component test intervals are optimized together.

To ensure high system reliability, redundancy allocation has been implemented on various industrial systems. For example, it has been recently applied on the renewable energy system design: in 2011, Xie and Billinton [24] proposed to minimize the total system cost (which are reliability-related, including the capital cost, maintenance and operation cost, and the customer interruption cost), through optimizing the number and types of wind turbine units installed at the multiple wind sites.

2.1.3 Component Assignment

In an industrial system, there are often interchangeable components, of differing quality and reliability, that can be allocated in the different positions of its functional logic and physical structure. For example, components with multiple functions can be interchangeable; identical components at different ages can be interchangeable. The overall system reliability can be improved by a proper assembly of such components into the required positions. In 1972 and 1974, Derman et al. [25] first formulated this problem in a parallel system and solved it with a method extended from the sum of products. Later on, the problem has been extended to series-parallel (and parallel-series) systems, consecutive k-out-of-n systems and general coherent systems. An overview on these research works can be found in [1].

Consider a representative BSSPS with m components to be assigned to k positions in the system. At each subsystem i, there are a set S_i of available positions. We have $k=\sum_{i=1}^{N}|S_i|$. Let $y_{hj}=1$ denote the component j that is assigned to position h and 0 otherwise. The assignment of the components can be represented by the vector $\boldsymbol{y}=(y_{11},\ldots,y_{1m},\ldots,y_{k1}\ldots,y_{km})$. Then the optimal assignment problem is formulated as follows:

$$\max R=\prod_{i=1}^{N}\left(1-\prod_{h\in S_i}\prod_{j=1}^{m}\left(1-r_{hj}\right)^{y_{hj}}\right) \tag{2.3a}$$

$$\text{s.t.}\sum_{j=1}^{m}y_{hj}=1,\quad h=1,2,\ldots,k \tag{2.3b}$$

$$\sum_{h=1}^{k}y_{hj}\leq 1;\ y_{hj}\in 0,1\ \ j=1,2,\ldots,m \tag{2.3c}$$

where r_{hj} is the reliability of component j assigned to position h.

In recent studies by Lin and Yeh [26,27], the component assignment problem is extended to computer, communication, and power networks. The generic network reliability is defined as the probability that the network can transmit d units of commodities from an origin to a specific destination. The power network expansion problem studied by Cadini, et al., [28] is similar to a component assignment problem in the sense that the optimal expansion solution seeks to add (i.e., allocate) transmission lines in proper locations so as to maximize the network reliability.

2.1.4 Maintenance and Testing

The engineered safety systems, e.g., the high-pressure injection system (HPIS) in a nuclear power plant (NPP), are usually under periodical tests and maintenances to reveal and repair the failures that may have occurred since the previous inspection. In such systems, a period of downtime can be caused by either failure or testing and maintenance, which makes the system unavailable. To ensure system safety, the system loss or cost due to the downtime has to be minimized. In the late 1960s and early 1970s, Jacobs [29] and Hirsch [30] attempted to find the best test intervals that minimize the time-average unavailability. In 1995, Vaurio [31] considered a more general formulation, which includes the cost minimization together with an unavailability constraint to search for the optimal test and maintenance intervals. A trade-off between the cost of system testing and unavailability is considered, because frequent testing usually increases the cost whereas infrequent testing usually leads to high unavailability. In the works by Munõz, et al. [32], Martorell, et al. [33], and Busacca, et al. [34], different methods were developed to find the optimal test or maintenance intervals so as to minimize system cost and unavailability.

The system unavailability is often defined on the basis of the minimal cut sets, which was found as a result of the fault tree analysis of the system. To give an example, the system unavailability has the following approximate expression, as reported in [35]:

$$U(\boldsymbol{y}) \approx \sum_{j=1}^{N_m} \prod_{i=1}^{n_j} u_i^j(\boldsymbol{y}), \tag{2.4}$$

where $\boldsymbol{y}$ is the vector of decision variables that governs the system availability characteristics and maintainability activities, N_m is the total number of minimal cut sets, n_j is the number of basic events (i.e. the number of components in binary state setting) relevant to the j-th minimal cut set, and $u_i^j(\boldsymbol{y})$ is the unavailability associated with the i-th component belonging to the j-th minimal cut.

As to the unavailability of the i-th generic component, several models are available in the literature to account for different contributions from failure on demand, maintenance, etc. Below is the original model presented in [31]:

$$u_i(\boldsymbol{y}_i) = \rho_i + \frac{1}{2}\lambda_i T_i + \frac{t_i}{T_i}, \tag{2.5}$$

where ρ_i is the probability of failure on demand, λ_i is the failure rate, T_i is the test interval and t_i is the mean downtime due to a test or maintenance carried out within T_i. In this case, $\boldsymbol{y}_i = (\rho_i, \lambda_i, T_i, t_i)$. Each of these parameters can, in turn, be a function of other parameters related to the causes of unavailability. This formulation implies that any unavailability contribution can be represented as a term being (1) independent of T_i, (2) proportional to T_i, or (3) inversely proportional to T_i [31]. Later on, this model was extended in various studies [34,36].

The cost function of the system is a sum of the cost of the individual components that constitute the system

$$C(\boldsymbol{y}) = \sum_{i=1}^{N_c} c_i(\boldsymbol{y}), \tag{2.6}$$

where N_c is the total number of components in the system and $c_i(\boldsymbol{y})$ is the cost allocated to each component. Typically, the component cost is made up of two major contributors: the costs of test and maintenance and the costs of the consequences of the failures [34]. Besides unavailability and cost, other objectives are considered by different researchers. For examples, Čepin and Mavko [37] consider the time-dependent failure probability of the system. Martorell, et al. [38] consider risk, reliability, and maintainability. The problem is essentially multi-objective. The formulation and solution approaches to multi-objective optimization will be presented in Chapter 10.

In more recent studies, a number of maintenance and test parameters, different from the previous elements of $\boldsymbol{y}$, were used as the decision variables. In [39], Wang and Pham considered the number of imperfect maintenance actions and the length of the initial imperfect maintenance interval. Yang and Chang [40] regarded the type of maintenance actions (e.g. no maintenance, minor and major maintenance) as the decision variables. Liu, et al. [41] introduced a maintenance threshold level. Khatab, et al. [42] considered the reliability threshold together with the number of preventive maintenance actions as decision variables.

In the following subsections, we will introduce three commonly used policies for component maintenance and replacements [43].

2.1.4.1 Age Replacement Policy

Under an age replacement policy [44], a component is replaced after a constant time T since its installation, or, at its failure, whichever occurs first. In general, we take the age replacement policy under the following assumptions: 1) the failures are instantly detected; 2) each failed component should be replaced with a new one; and, 3) the replacement time is negligible; 4) the failure time $X_k\ (k=1,2,\ldots)$ of each component is independent and has an identical distribution $F(t)=\Pr(X_k \le t)$, with mean μ.

Now, assume a new component is installed at time $t=0$, an age replacement procedure generates a renewal process, and, let $\{X_k\}_{k=1}^{\infty}$ be the failure times of successive operating components, with $Z_k=\min\{X_k,T\}$. Then, $\{Z_k\}_{k=1}^{\infty}$ represents the length of the intervals between each replacements k, which may caused by either failures or planned replacements, and we have

$$\Pr\{Z_k \le t\}=\begin{cases} F(t), & \text{for}\, t<T \\ 1, & \text{for } t \ge T \end{cases}.$$

We consider the problem of minimizing the expected cost per unit of time for an infinite time span. We introduce the following costs: c_1 the failure cost, and $c_2(<c_1)$ the replacement cost. Let $N_1(t)$ denote the number of failures during $(0,t]$ and $N_2(t)$ denote the number of replacements with a working component during $=(0,t]$. Then, the expected cost during $(0,t]$ is given by

$$\hat{C}(t)=c_1E(N_1(t))+c_2E(N_2(t)).$$

We call the time interval from one replacement to the next replacement as one cycle. Thus, the expected cost per unit of time for an infinite time span is

$$C(T)=\lim_{t\to\infty}\frac{\hat{C}(t)}{t}=\frac{\text{expected cost of one cycle}}{\text{mean time of one cycle}}.$$

We call $C(T)$ the expected cost rate, and, generally adopt it as the objective function of an optimization problem. When we set a planned replacement at time T with failure time X, the expected cost of one cycle is

$$c_1\Pr(X\le T)+c_2\Pr(X>T)=c_1F(T)+c_2\bar{F}(T).$$

The mean time of one cycle is

$$\int_0^T t dPr(X\le t)+TPr(X>T)=\int_0^T t dF(t)+T\bar{F}(T)=\int_0^T \bar{F}(t)dt.$$

Thus, the expected cost rate is

$$C(T)=\frac{c_1F(T)+c_2\bar{F}(T)}{\int_0^T\bar{F}(t)dt}.$$

If $T=\infty$, then the policy corresponds to the replacement only at failure, and the expected cost rate is

$$C(\infty)=\lim_{t\to\infty}C(T)=\frac{c_1}{\mu}.$$

2.1.4.2 Periodic Replacement Policy

When a reliability system is very complex and large-scale, one should allow for only minimal repair at each failure, and only performs the replacement actions periodically. We call such policy as the periodic replacement with minimal repair at failures [45], which is introduced as follows: Suppose the failure times of a component have a density function $f(t)$ and a cumulative distribution $F(t)$ with finite mean μ and failure rate $h(t)=f(t)/\bar{F}(t)$, where $\bar{F}(t)=1-F(t)$. Consider one cycle with constant time T from the planned replacement to the next one. Let c_1 be the cost of minimal repair and c_2 be the cost of the planned replacement. Then, the expected cost of one cycle is

$$c_1E(N_1(t))+c_2E(N_2(t))=c_1H(T)+c_2$$

which is because the expected number of failures during one cycle is $E(N_1(t))=\int_0^T h(t)dt$ $= H(T)$, as proven by Theorem 4.1 in [46]. Therefore, the expected cost rate is

$$C(T)=\frac{1}{T}\left[c_1H(T)+c_2\right].$$

2.1.4.3 Block Replacement Policy

Under the block replacement policy [47], two types of actions are considered: the preventive replacement that occurs at fixed intervals of time regardless of the state of the component and the failure replacement when the component fails. The objective is to determine the optimal time interval to replace and optimize the expected cost rate. Suppose the replacement is conducted at time $kT\ (k=1,2,\ldots)$ and each component has an identical time to failure cdf $F(t)$ with mean μ. $F^{(n)}(t)\ (n=1,2,\ldots)$ is the n-fold convolution of $F(t)$. Consider one cycle with constant time T from the planned replacement to the next one. Let c_1 be the cost of replacement of a failed component and c_2 be the cost of the planned replacement. Then, because the expected number of failed components during one cycle is $M(T)=\sum_{n=1}^{\infty}F^{(n)}(T)$ from (1.19) in [46], the expected cost in one cycle is

$$c_1E(N_1(t))+c_2E(N_2(t))=c_1M(T)+c_2.$$

Therefore, the expected cost rate is

$$C(T)=\frac{1}{T}\left[c_1M(T)+c_2\right].$$

If a component is replaced only at failure, i.e. $T = \infty$, then, we have $\lim_{T\to\infty} M(T)/T = 1/\mu$ and the expected cost rate is

$$C(\infty) = \lim_{T\to\infty} C(T) = \frac{c_1}{\mu}.$$

2.2 Optimization Methods

In this section, we provide basics of commonly-used modelling and solution schemes for reliability optimization problems.

2.2.1 Mathematical Programming

Dantzig and Thapa [48] define mathematical programming (MP) as "...the branch of mathematics dealing with techniques for maximizing or minimizing an objective function subject to linear, non-linear, and integer constraints on the variables." MP utilizes mathematical theory and computational solutions to assist in decision making, usually regarding the best use of (scarce) resources. MP can be written as

$$\max(\text{or } \min) f(\boldsymbol{x}) \tag{2.7a}$$

$$\text{s.t. } g_j(\boldsymbol{x}) \geq 0, j = 1,2,\ldots,m \tag{2.7b}$$

$$h_j(\boldsymbol{x}) = 0, j = m+1, m+2,\ldots,p \tag{2.7c}$$

where $\boldsymbol{x}$ is the decision variable with n dimensions. $\mathbf{g}(\boldsymbol{x}) = (g_1(\boldsymbol{x}), g_2(\boldsymbol{x}),\ldots, g_m(\boldsymbol{x}))$ are the inequality constraints and $\boldsymbol{h}(\boldsymbol{x}) = (h_{m+1}(\boldsymbol{x}), h_2(\boldsymbol{x}),\ldots, h_p(\boldsymbol{x}))$ are the equality constraints. The set of the elements of the definition space that satisfy all the constraints is called the feasible region. The decision makers assess the quality of the possible alternative solutions with respect to a given criterion function $f(\boldsymbol{x})$, which is called objective function.

MP can be classified in several ways, such as based on the nature of problem or equations, the type of decision variables, etc. Based on the structure of a problem, MP involves

- models with linear functions, i.e. linear programming (LP)
- models with only integer variables, i.e. integer programming (IP)
- models with more general functions, i.e. non-linear programming (NLP)
- models with continuous and discrete variables, i.e. mixed integer programming (MIP)
- models with random parameters, i.e. stochastic programming (SP)

Classical MP problems include set coverage problem, knapsack problem, and traveling salesman problem, etc. Along with the development of MP research, many numerical solution methods, such as dynamic programming (DP), branch and bound (B&B), and column generation (CG) has been proposed for solve MP problems, each take advantage of specific problem structures. We introduce these methods in the following subsections.

2.2.1.1 Branch-and-Bound (B&B)

The B&B method [50] is a divide-and-conquer method based on an efficient enumeration of the possible feasible solutions. The principle of B&B is to divide the solution space into disjoint subsets, which are denoted by the nodes of the branching tree. Then, the algorithm explores the other nodes of the branching tree under a given strategy. To avoid exploring the entire branching tree, the algorithm assesses the node before branching a new node. The best solution which might be found in the associated subtree is compared with the best solution that has currently been obtained. If the best solution under the node is worse than the current best solution, the subtree is discarded. Otherwise, the node is branched and the above operations are repeated. The application of B&B is widely used in IP or MIP with binary variables. The illustration of B&B used for binary variables is shown in Figure 2.1.

2.2.1.2 Dynamic Programming (DP)

DP has a rich and diverse history in the field of mathematics [51]. It is an optimization method based on the principle defined by Bellman: "An optimal policy has the property that whatever the initial state and initial decision are, the remaining decisions must constitute an optimal policy with regard to the state resulting from the first decision." It can be summarized as follows: A multi-stage problem can be decomposed into a sequence of interrelated one-stage problems. The optimal solution for the multistage problem must consist of optimal policies for its substage problem.

To introduce DP to solve the multi-stage problems in a general framework, we conduct a simple example of a shortest path problem shown in Figure 2.2. We consider a directed graph $D=(V,A)$ with the arc distance c_e for arc $e \in A$. The problem is to find the shortest path from the source node s to the sink node t. We observe that if the shortest path from s to t passes through the node w, then the subpaths (s,w) and (w,t) must be the shortest paths from s to w and w to t, respectively. Let $d(v)$ denote the length of the shortest path from s to v. Then

$$d(v)=\min_{w\in V^-(v)}\{d(w)+c_{wv}\},$$

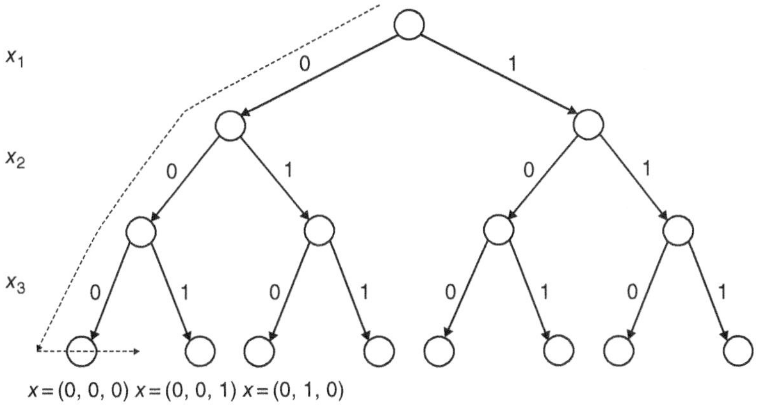

Figure 2.1 Example for the search tree in B&B.

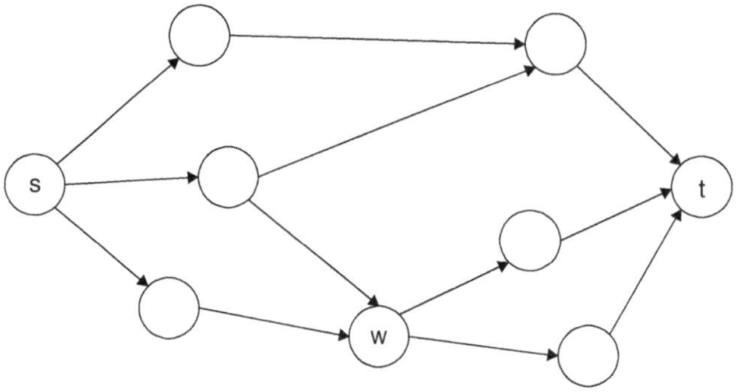

Figure 2.2 Shortest $s-t$ path.

where $V^{-}(v)$ denotes the processor set in $G(V,A)$ of node v. That is, the shortest path from s to v is the shortest path from s to the neighbor of v and then to v.

2.2.1.3 Column Generation (CG)

CG [52] refers to an algorithm to solve LP problems when there are a huge number of variables compared to the number of constraints. Instead of enumeration, the simplex algorithm is used in CG to decide whether the current best solution is optimal.

To illustrate CG solving a LP problem, we consider a simple example, which refers to the following problem:

$$(\text{LP}) \quad z = \max\left\{\sum_{k=1}^{K} c^k x^k : \sum_{k}^{K} A^k x^k = b,\ x^k \in X^k, k = 1,\ldots,K\right\} \tag{2.8}$$

where $X^k = \left\{x^k \in R_+^{n_k} : D^k x^k \le d_k\right\}$ for $k = 1,\ldots,K$. Assuming that all sets X^k consist of a huge but finite set of points $\left\{x^{k,t}\right\}_{t=1}^{T_k}$, we have

$$X^k = \{x^k \in R^{n_k} : x^k = \sum_{t=1}^{T_k} \lambda_{k,t} x^{k,t}, \sum_{t=1}^{T_k} \lambda_{k,t} = 1,\ \lambda_{k,t} \ge 0, \forall t = 1,\ldots,T_k.$$

Now, we replace x^k with $\sum_{t=1}^{T_k} \lambda_{k,t} x^{k,t}$, leading to an equivalent LP master problem (LPM):

$$z^{LPM} = \max \sum_{k=1}^{K} \sum_{t=1}^{T_k} \left(c^k x^{k,t}\right) \lambda_{k,t} \tag{2.9a}$$

$$(\text{LPM}) \quad \sum_{k=1}^{K} \sum_{t=1}^{T_k} \left(A^k x^{k,t}\right) \lambda_{k,t} = b \tag{2.9b}$$

$$\sum_{t=1}^{T_k} \lambda_{k,t} = 1,\ \forall k = 1,\ldots,K \tag{2.9c}$$

$$\lambda_{k,t} \ge 0,\ \forall t = 1,\ldots,T_k,\ k = 1,\ldots,K \tag{2.9d}$$

In the following, we present the CG procedure to solve the above LPM.

Initialization. $\left(\left(c^k x\right)^T, \left(A^k x\right)^T, e_k^T\right)^T$ is a column (vector) for LPM and for each $x \in X^k$.

We assume a subset of columns is known as the initialization (at least one for each X^k), providing a feasible restricted LPM (RLPM):

$$\tilde{z}^{LPM} = \max \tilde{c}\tilde{\lambda} \tag{2.10a}$$

$$(\text{RLPM}) \quad \tilde{\lambda} = \tilde{b} \tag{2.10b}$$

$$\tilde{\lambda} \geq 0 \tag{2.10c}$$

where $\tilde{c}$ and $\tilde{A}$ are the sub-matrices of the original parameter matrix with the initialized columns. Solving RLPM provides an optimal primal solution $\tilde{\lambda}^*$ and an optimal dual solution $(\pi, \mu) \in R^m \times R^K$ where π represents the dual variables associated with the joint constraints (2.9b), and μ represents the dual variables for the constraints (2.9c).

Primal Feasibility. Any feasible solution of RLPM is feasible for LPM. Because $\tilde{\lambda}^*$ is a feasible solution for LPM, $\tilde{c}\tilde{\lambda}^*$ will give a lower bound for LPM, meaning - $\tilde{z}^{LPM} = \tilde{c}\tilde{\lambda}^* = \sum_{i=1}^{m} \pi_i b_i + \sum_{k=1}^{K} \mu_k \leq z^{LPM}$.

Optimality Check for LPM. We need to check whether (π, μ) is dual-feasible for LPM. This means to check whether the reduced price $c^k x - \pi A^k x - \mu_k$ of each column $\left(\left(c^k x\right)^T, \left(A^k x\right)^T, e_k^T\right)^T$ for each k and for each $x \in X^k$ is no more than zero. Rather than checking all possible points in X^k enumeration, we treat all points implicitly by solving the following subproblem:

$$\varsigma_k = \max\left\{\left(c^k - \pi A^k\right)x - \mu_k : x \in X^k\right\}$$

Generating a New Column. If $\varsigma_k > 0$ is for some k, the column corresponding to the optimal solution $\tilde{x}^k$ of the subproblem will have a positive reduced price. We introduce the column $\left(\left(c^k \tilde{x}^k\right)^T, \left(A^k \tilde{x}^k\right)^T, e_k^T\right)^T$ into RLPM and re-optimize it until the stopping criterion is reached.

Stopping Criterion. If $\varsigma_k \leq 0$ is for $k = 1, \ldots, K$, the solution (π, μ) will be dual-feasible for LPM; therefore, $\sum_{i=1}^{m} \pi_i b_i + \sum_{k=1}^{K} \mu_k$ is an upper bound of LPM, meaning $z^{LPM} \leq \sum_{i=1}^{m} \pi_i b_i + \sum_{k=1}^{K} \mu_k$, and $\tilde{\lambda}$ is optimal for LPM.

2.2.2 Meta-heuristics

Meta-heuristic is a general search method used to solve complex combinatorial optimization problems. These problems are computationally challenging because solving them involves examining a huge number (usually exponential) of solutions made by the combination of values of the decision variables, and, evaluating the objective function in correspondence, to identify the optimal solution. For example, let us consider for the case that there are n jobs waiting to be completed by m machines. The processing time of the jobs and the processing power of machines are given. The goal is to schedule n jobs to m machines so as to minimize the makespan, which is the total length of the schedule that all jobs have finished processing. When the numbers of n and m are small, the problem is easily solved by enumerating all possible combinations; however, when the number of jobs and machines are large, the number of combinations increases exponentially, which makes the problem intractable by enumeration. These combinatorial problems suffering from the solutions explosion phenomenon are common in practical engineering where many parameters are involved, such as for planning a production process and designing a system. Therefore, it is important to devise intelligent algorithms to solve them in reasonable computational time. Moreover, the use of exact methods for the solution of these problems generally requires the problems be convex and linear, the objective of the optimization be differentiable and so on, which is often not the case in practice. There is, then, a great need for intelligent algorithms capable of solving these kinds of problems without having to pose restrictions on the mathematical properties.

Instead of enumerating all possible combinations of solutions, meta-heuristic methods construct the candidate solutions following certain policies. They gradually learn about the structure of the problem from the objective function values evaluated at the candidate solutions identified in the successive steps of the iterative search. The knowledge gained is used to construct new candidate solutions of improved quality, i.e. improved values of the objective function. Although these methods may miss the global optimal solution but, as a counterpart, they can often provide sub optimal results with good qualities, within an acceptable computational time. Moreover, meta-heuristic methods are derivative-free procedures, which do not require particular restrictions on the mathematical properties of the problems, such as convexity and differentiability. Therefore, the meta-heuristic methods can deal with a wide range of practical optimization problems including those non-convex, non-linear, and discrete ones.

In recent years, advanced meta heuristic methods which mimic behavioral policies in populations, have been widely proposed and studied. To name a few, it includes the cooperative behaviour of bees in the ABC algorithm, the social-spider behavior in the social spider optimization (SSO) algorithm, the social behavior of birds in the PSO algorithm, and the emulation of differential and conventional evolution of species in the differential evolution (DE) and genetic algorithm policies, etc.

2.2.2.1 Genetic Algorithm (GA)

GA is one of the most classic population-based meta-heuristic algorithms. Every solution provided by GA is coded into a chromosome where each decision variable is represented by a gene. GA first randomly generates an initial population; then, the 'selection', 'crossover', and 'mutation' operators are applied to modify the individuals of the population until the maximum number of generations is reached. GA utilizes the fitness function to evaluate the fitness of each individual in the population. To improve the quality of solutions, the crossover operator is used to generate new solutions whereby the solutions candidates are selected probabilistically, with a given selection mechanism, to generate the offsprings that enter the next population. The selection mechanism will more probably choose the solutions with better values of the objective function (called fitness in GA terminology). To avoid trapping into the local optimal solutions, the mutation operator is applied to pull out the population from the local region by randomly changing the genes of its individuals. The important steps of GA are described as follows:

- *Selection.* The selection of parents to generate successive populations in GA aims to select the fittest individuals (i.e. those with best values of the objective function) with highest probabilities. A number of selection mechanisms are available in literature [53]: roulette wheel, Boltzmann selection, tournament selection, rank selection, and steady-state selection.
- *Crossover.* After selecting the individuals, the crossover operator proceeds to generate individuals of the new population. The main task of the crossover operator is to combine two different individuals to generate a new one. There are different techniques for crossover operator in literature [54], such as single-point and double-point crossover, three parents crossover, cycle crossover, order crossover, masked crossover.
- *Mutation.* The last main operator is mutation, the application in which certain genes of individuals are altered with a small mutation rate. The mutation operator prevents the population from remaining trapped in local optima. Some of the popular mutation operators [55] are the following: power mutation, shrink, Gaussian, uniform.

2.2.2.2 Differential Evolution (DE)

DE [56] is also a population-based algorithm. Unlike GA, DE perturbs the current individuals with the scaled differences of randomly selected and distinct individuals. This property allows DE to perform fewer mathematical operations than GA and other algorithms, hence requiring reduced execution time compared to other algorithms. In DE, the individual solutions are called parameter vectors or genomes, like in GA. A parent vector from the current generation is called a target vector, a mutant vector obtained after the mutation operation is called a donor vector, and an offspring individual constructed by recombining the donor vector with the target vector is called a trial vector. The important steps of DE are described as follows.

- *Mutation.* Unlike GA, the mutation operator in DE is performed on all target vectors (parameter vectors, individuals) at every iteration.

- *Crossover*. The crossover operator is applied after generating the donor vector by means of the mutation operator. This operator is utilized to enhance the diversity of the population by exchanging the components of the donor vector with the target vector, to generate the trial vector. Exponential and binomial crossover methods are typically used in DE.
- *Selection*. The selection operator decides whether the trial vector or target vector is selected to enter the successive generation. The vector with best fitness value is selected to ensure the population never deteriorates.

2.2.2.3 Particle Swam Optimization (PSO)

PSO [57] is inspired by the social behavior of organisms within large groups, such as birds, fish, and even humans. This algorithm emulates the interaction between members to share information. Its main advantage is its fast convergence compared with many other global optimization algorithms. PSO consists of a swarm of particles, whose trajectories are adjusted by a stochastic term and a deterministic term. Each particle is influenced by its best reached position and the best position reached by a particle member of the group but tends to move randomly. A particle i is denoted by its velocity vector v_i and its position vector x_i. In every iteration, each particle adjusts its position according to its new velocity:

$$v_i^{t+1} = wv_i^t + c_1 r_1 \left(xBest_i^t - x_i^t \right) + c_2 r_2 \left(gBest_i^t - x_i^t \right)$$

$$x_i^{t+1} = x_i^t + v_i^t \cdot t$$

where *xBest* and *gBest* are the best positions of the particle and group, respectively. The parameter w, c_1, c_2, r_1, r_2 denotes the weight with two positive constants and two random parameters, respectively.

2.3 Exercises

1 A unit wears out according to a normal distribution with a mean of 1,000,000 cycles and standard deviation of 100,000 cycles. The cost of preventive replacement is \$50 and that of the corrective replacement is \$100. Assume the preventive replacements can be performed at discrete time intervals equivalent to 100,000 cycles per interval. Please determine the optimal preventive replacement interval by using block replacement policy.

2 Solve the Problem 1 by using the age replacement policy. Determine the optimal preventive replacement interval.

3 A system is found to exhibit a constant failure rate of 5×10^{-5} failures per hour. The system is repaired upon failure and then, returned to its original condition. What is the expected number of failures after two years of operation?

4 Use any exact method to solve the following redundancy allocation problem, and, then, solve it by a heuristic algorithm. Compare the results of them.

$$\min \sum_{i=1}^{5} c_i x_i$$

$$\text{s.t.} \prod_{i=1}^{5} \left[1-(1-r_i)^{x_i}\right] \geq 0.9$$

$$x_i \in N_+, \forall i = 1,\ldots,5$$

where the parameters are set as

Subsystem i	1	2	3	4	5
r_i	0.93	0.89	0.91	0.88	0.92
c_i	1.3	2.2	3.1	2.5	2.9

(Solution: DP method for redundancy allocation problem proposed in [12]).

References

1 Kuo, W. and Prasad, V.R. (2000). An annotated overview of system-reliability optimization. *IEEE Transactions on Reliability* 49 (2): 176–187.

2 Kuo, W. and Wan, R. (2007). Recent advances in optimal reliability allocation. *IEEE Transactions on Systems, Man, and Cybernetics - Part A: Systems and Humans* 37 (2): 143–156.

3 Kulshrestha, D. and Gupta, M. (1973). Use of dynamic programming for reliability engineers. *IEEE Transactions on Reliability* 22 (4): 240–241.

4 Tillman, F.A., Kuo, W., and Hwang, C.-L. (1980). *Optimization of Systems Reliability*. Marcel Dekker.

5 Yun, W.Y., Park, G., and Han, Y.J. (Sep 2014). An optimal reliability and maintainability design of a searching system. *Communications in Statistics-Simulation and Computation* 43 (8): 1959–1978. doi:10.1080/03610918.2013.815771.

6 Azaron, A., Perkgoz, C., Katagiri, H., Kato, K., and Sakawa, M. (May 2009). Multi-objective reliability optimization for dissimilar-unit cold-standby systems using a genetic algorithm. *Computers & Operations Research* 36 (5): 1562–1571. doi:10.1016/j.cor.2008.02.017.

7 Huang, H.Z., Qu, J., and Zuo, M.J. (2009). Genetic-algorithm-based optimal apportionment of reliability and redundancy under multiple objectives. *IIE Transactions* 41 (4): 287–298. doi:10.1080/07408170802322994.

8 Ghare, P. and Taylor, R. (1969). Optimal redundancy for reliability in series systems. *Operations Research* 17 (5): 838–847.
9 Beraha, D. and Misra, K. (1974). Reliability optimization through random search algorithm. *Microelectronics Reliability* 13 (4): 295–297.
10 Chern, M.-S. (1992). On the computational complexity of reliability redundancy allocation in a series system. *Operations Research Letters* 11 (5): 309–315.
11 Sup, S.C. and Kwon, C.Y. (1999). Branch-and-bound redundancy optimization for a series system with multiple-choice constraints. *IEEE Transactions on Reliability* 48 (2): 108–117.
12 Yalaoui, A., Châtelet, E., and Chu, C. (2005). A new dynamic programming method for reliability & redundancy allocation in a parallel-series system. *IEEE Transactions on Reliability* 54 (2): 254–261.
13 Zia, L. and Coit, D.W. (Dec 2010). Redundancy allocation for series-parallel systems using a column generation approach. *IEEE Transactions on Reliability* 59 (4): 706–717.
14 Ouzineb, M., Nourelfath, M., and Gendreau, M. (2010). An efficient heuristic for reliability design optimization problems. *Computers & Operations Research* 37 (2): 223–235.
15 Ushakov, I. (1987). Optimal standby problems and a universal generating function. *Soviet Journal of Computer and Systems Sciences* 25 (4): 79–82.
16 Ushakov, I. (1986). Universal generating function. *Soviet Journal of Computer and System Sciences* 24 (5): 37–49.
17 Levitin, G., Lisnianski, A., BenHaim, H., and Elmakis, D. (Jun 1998). Redundancy optimization for series-parallel multi-state systems. *IEEE Transactions on Reliability* 47 (2): 165–172.
18 Massim, Y., Zeblah, A., Meziane, R., Benguediab, M., and Ghouraf, A. (2005). Optimal design and reliability evaluation of multi-state series-parallel power systems. *Nonlinear Dynamics* 40 (4): 309–321.
19 Wang, Y. and Li, L. (Mar 2012). Heterogeneous redundancy allocation for series-parallel multi-state systems using hybrid particle swarm optimization and local search. *IEEE Transactions on Systems, Man, and Cybernetics - Part A: Systems and Humans* 42 (2): 464–474.
20 Hsieh, T.-J. and Yeh, W.-C. (2012). Penalty guided bees search for redundancy allocation problems with a mix of components in series–parallel systems. *Computers & Operations Research* 39 (11): 2688–2704.
21 Liu, Y., Huang, H.Z., Wang, Z.L., Li, Y.F., and Yang, Y.J. (Jun 2013). A joint redundancy and imperfect maintenance strategy optimization for multi-state systems. *IEEE Transactions on Reliability* 62 (2): 368–378. doi:10.1109/tr.2013.2259193.
22 Lins, I.D. and Droguett, E.L. (Jan 2011). Redundancy allocation problems considering systems with imperfect repairs using multi-objective genetic algorithms and discrete event simulation. *Simulation Modelling Practice and Theory* 19 (1): 362–381. doi:10.1016/j.simpat.2010.07.010.
23 Zio, E. and Podofillini, L. (Oct 2007). Importance measures and genetic algorithms for designing a risk-informed optimally balanced system. *Reliability Engineering & System Safety* 92 (10): 1435–1447. doi:10.1016/j.ress.2006.09.011.

24 Xie, K.G. and Billinton, R. (Mar 2011). Determination of the optimum capacity and type of wind turbine generators in a power system considering reliability and cost. *IEEE Transactions on Energy Conversion* 26 (1): 227–234. doi:10.1109/tec.2010.2082131.

25 Derman, C., Lieberman, G.J., and Ross, S.M. (1972). On optimal assembly of systems. *Naval Research Logistics Quarterly* 19 (4): 569–574.

26 Lin, Y.K. and Yeh, C.T. (Aug 2011). Multistate components assignment problem with optimal network reliability subject to assignment budget. *Applied Mathematics and Computation* 217 (24): 10074–10086. doi:10.1016/j.amc.2011.05.001.

27 Lin, Y.K. and Yeh, C.T. (May 2012). Multi-objective optimization for stochastic computer networks using NSGA-II and TOPSIS. *European Journal of Operational Research* 218 (3): 735–746. doi:10.1016/j.ejor.2011.11.028.

28 Cadini, F., Zio, E., and Petrescu, C.A. (Mar 2010). Optimal expansion of an existing electrical power transmission network by multi-objective genetic algorithms. *Reliability Engineering & System Safety* 95 (3): 173–181. doi:10.1016/j.ress.2009.09.007.

29 Jacobs, I. (1968). Reliability of engineered safety features as a function of testing frequency. *Nulcear Safety* 9: 302–312.

30 Hirsch, H.M. (1971). Setting test intervals and allowable bypass times as a function of protection system goals. *IEEE Transactions on Nuclear Science* 18 (1): 488–494.

31 Vaurio, J. (1995). Optimization of test and maintenance intervals based on risk and cost. *Reliability Engineering & System Safety* 49 (1): 23–36.

32 Munoz, A., Martorell, S., and Serradell, V. (1997). Genetic algorithms in optimizing surveillance and maintenance of components. *Reliability Engineering & System Safety* 57 (2): 107–120.

33 Martorell, S., Villamizar, M., Carlos, S., and Sanchez, A. (Dec 2010). Maintenance modeling and optimization integrating human and material resources. *Reliability Engineering & System Safety* 95 (12): 1293–1299. doi:10.1016/j.ress.2010.06.006.

34 Busacca, P.G., Marseguerra, M., and Zio, E. (2001). Multiobjective optimization by genetic algorithms: Application to safety systems. *Reliability Engineering & System Safety* 72 (1): 59–74.

35 Henley, E.J. and Kumamoto, H. (1981). *Reliability Engineering and Risk Assessment*. Englewood Cliffs (NJ): Prentice-Hall.

36 Martorell, S., Sánchez, A., Carlos, S.A., and Serradell, V. (2002). Simultaneous and multi-criteria optimization of TS requirements and maintenance at NPPs. *Annals of Nuclear Energy* 29 (2): 147–168.

37 Čepin, M. and Mavko, B. (1997). Probabilistic safety assessment improves surveillance requirements in technical specifications. *Reliability Engineering & System Safety* 56 (1): 69–77.

38 Martorell, S. et al. (2005). RAMS+C informed decision-making with application to multi-objective optimization of technical specifications and maintenance using genetic algorithms. *Reliability Engineering & System Safety* 87 (1): 65–75.

39 Wang, L., Chu, J., and Mao, W. (2009). A condition-based replacement and spare provisioning policy for deteriorating systems with uncertain deterioration to failure. *European Journal of Operational Research* 194 (1): 184–205.

40 Yang, F. and Chang, C.S. (Nov 2009). Multiobjective evolutionary optimization of maintenance schedules and extents for composite power systems. *IEEE Transactions on Power Systems* 24 (4): 1694–1702. doi:10.1109/tpwrs.2009.2030354.

41 Liu, X., Li, J.R., Al-Khalifa, K.N., Hamouda, A.S., Coit, D.W., and Elsayed, E.A. (Apr 2013). Condition-based maintenance for continuously monitored degrading systems with multiple failure modes. *IIE Transactions* 45 (4): 422–435. doi:10.1080/0740817x.2012.690930.

42 Khatab, A., Ait-Kadi, D., and Rezg, N. (2014). Availability optimisation for stochastic degrading systems under imperfect preventive maintenance. *International Journal of Production Research* 52 (14): 4132–4141. doi:10.1080/00207543.2013.835499.

43 Elsayed, E.A. (2020). *Reliability Engineering*. John Wiley & Sons.

44 Badía, F., Berrade, M., and Lee, H. (2020). An study of cost effective maintenance policies: Age replacement versus replacement after N minimal repairs. *Reliability Engineering System Safety* 201: 106949.

45 Dong, W., Liu, S., and Du, Y. (2019). Optimal periodic maintenance policies for a parallel redundant system with component dependencies. *Computers Industrial Engineering* 138: 106133.

46 Nakagawa, T. (2006). *Maintenance Theory of Reliability*. Springer Science & Business Media.

47 Ke, H. and Yao, K. (2016). Block replacement policy with uncertain lifetimes. *Reliability Engineering & System Safety* 148: 119–124.

48 Dantzig, G.B. and Thapa, M.N. (1997). The Linear Programming Problem. In: *Linear Programming: 1: Introduction*, 1–33.

49 Vanderbei, R.J. (2015). *Linear Programming*. Springer.

50 Lawler, E.L. and Wood, D.E. (1966). Branch-and-bound methods: A survey. *Operations Research* 14 (4): 699–719.

51 Bellman, R.E. and Dreyfus, S.E. (2015). *Applied Dynamic Programming*. Princeton university press.

52 Desaulniers, G., Desrosiers, J., and Solomon, M.M. (2006). *Column Generation*. Springer Science & Business Media.

53 Shukla, A., Pandey, H.M., and Mehrotra, D., "Comparative review of selection techniques in genetic algorithm," in *2015 international conference on futuristic trends on computational analysis and knowledge management (ABLAZE)*, 2015: IEEE, pp. 515–519.

54 Kora, P. and Yadlapalli, P. (2017). Crossover operators in genetic algorithms: A review. *International Journal of Computer Applications* 162 (10).

55 Deep, K. and Thakur, M. (2007). A new mutation operator for real coded genetic algorithms. *Applied Mathematics Computation* 193 (1): 211–230.

56 Price, K.V. (2013). Differential evolution. In: Ivan ZelinkaVáclav SnášelAjith Abraham *Handbook of Optimization*. Springer, 187–214.

57 Clerc, M. (2010). *Particle Swarm Optimization*. John Wiley & Sons.

Part II

Reliability Techniques

3

Multi-State Systems (MSSs)

Conventional approaches typically assume that the components and the system have two states: perfect working and complete failure [1]. However, many engineering systems can carry on their intended tasks with various levels of performance, with the components being partially functioning. Let us take an offshore gas pipeline network as an example. The network has a compressor subsystem composed of multiple compressors, and, the state of the subsystem is defined as its capacity for gas transportation. Thus, the subsystem state is a function of the capacities of the functioning compressors. Thus, once certain compressors fail or under maintenance actions, the subsystem state shall change, and it could take multiple values in long run [2]. Another example is a wind turbine system. The generator component can produce power less efficiently due to the failure of the anemometer, which can lead to inaccurate readings of the wind speed on which to base the blades adjustments [3]. For such cases, the multi-state models (MSMs) offer higher flexibility and a more precise approximation than the binary state models in the modelling of the system state distribution and its real-world dynamics.

The first attempts to investigate MSMs appeared as theoretical studies by Barlow and Wu [4], El-Neweihi, et al. [5] and Ross [6] in the late 1970s, followed by the independent works by Griffith [7], Block and Savits [8], Butler [9], Natvig [10] and Ebrahimi [11] in the early 1980s. These works extended the theory of binary coherent structures to multi-state components and lay the foundation of multi-state theory by properly defining the concepts of multi-state monotone system (MSMS), multi-state coherent system (MSCS) (a special case of the former), minimal path vectors (MPVs), and minimal cut vectors (MCVs).

3.1 Classical Multi-state Models

In classical model, the component and system have the same set of possible working states. For a multi-state system (MSS) of n components, let $i \in \{1,\ldots,n\}$ denote the component index and $S = \{0,\ldots,M\}$ denote the state set of each component and the system, where M represents the state of perfect functioning, 0 represents the state of complete failure, and the other intermediate numbers represent the partially functioning states. Let $x_i \in S$ denote the state of component i; then, $\boldsymbol{x} = (x_1,\ldots,x_n)$ denotes the component

Reliability Analysis, Safety Assessment and Optimization: Methods and Applications in Energy Systems and Other Applications, First Edition. Enrico Zio and Yan-Fu Li.

state vector and $\boldsymbol{S} = S^n = \{\boldsymbol{x} \mid 0 \le x_i \le M, \text{ for } i = 1,\ldots,n\}$ denotes the component state space. For any two state vectors $\boldsymbol{x}$ and $\boldsymbol{y}$, $x < y$implies that $x_i \le y_i$ for all $i \in \{1,\ldots,n\}$ and $x_i < y_i$ for at least one $i \in \{1,\ldots,n\}$. The system state ϕ taking values from S is essentially a deterministic function of x, i.e. $\phi = \phi(\boldsymbol{x})$. The function $\phi(\cdot)$ is called the structure function of the system. The following assumptions concerning ϕ characterize different types of MSS:

i ϕ is non-decreasing, i.e. if $x \le y$, then $\phi(x) \le \phi(y)$;
ii $\phi(0) = 0$ and $\phi(\boldsymbol{M}) = M$ where $0 = (0,\ldots,0)$ and $\boldsymbol{M} = (M,\ldots,M)$;
iii for all $j \in \{0,\ldots,M\}$, $\phi(j) = j$ where $j = (j,\ldots,j)$; and
iv forall $i \in \{1,\ldots,n\}$ andall $j \in \{0,\ldots,M\}$,thereexistsavector $(\cdot,x) = (x_1,\ldots,x_{i-1},\cdot,x_{i+1},\ldots,x_n)$ such that for $k \ne j$, we have $\phi(j_i,x) = j$ and $\phi(k_i,x) \ne j$.That is, the system contains no irrelevant components.

The system with ϕ satisfying assumptions *i* and *ii* is called an MSMS, whereas the system with ϕ satisfying all the above assumptions is called an MSCS.

The definitions of MPV and MCV are key concepts in classical MSS theory:

- MPV:A vector$x$ is an MPV to level j if $\phi(x) \ge j$; and for any$y < x$, itimplies $\phi(y) < j$
- MCV:A vector$x$ is an MCV to level j if$\phi(x) < j$; for any$y > x$, itimplies $\phi(y) \ge j$

For a system under random setting, let a random variable X_i denote the state of component i; then, $p_{ij} = \Pr(X_i = j)$ is the probability of component i being at state j. Clearly, we have $\sum_{j=0}^{M} p_{ij} = 1$ for any i. Let random vector $X = (X_1,\ldots,X_n)$ denote the state of all components where $X_1,\ldots,X_n$ are assumed to be statistically mutually independent. Then, $\phi(X)$ is the random variable representing the system state, with $p_j = \Pr(\phi(X) = j)$.

The system reliability with respect to state j is defined as $R_\phi^j = \Pr(\phi(X) \ge j)$, i.e. the probability of occupying a state higher than j assuming j as lower threshold. Let $y_k^j = (y_{1k}^j,\ldots,y_{nk}^j)$ for $k \in \{1,\ldots,n_\phi^j\}$ (where n_ϕ^j is the total number of MPVs to state j) denote the k-th MPV to state j. Let $z_l^j = (z_{1l}^j,\ldots,z_{nl}^j)$ for $l \in \{1,\ldots,m_\phi^j\}$ (where m_ϕ^j is the total number of MCVs to state j) denote the l-th MCV to state j. Let A_k^j denote the event that $X \ge y_k^j$ and B_l^j denote the event that $X \le z_l^j$. Then

$$R_\phi^j = \Pr\left(\bigcup_{k=1}^{n_\phi^j} A_k^j\right) = 1 - \Pr\left(\bigcup_{l=1}^{m_\phi^j} B_l^j\right). \tag{3.1}$$

We have $\Pr(A_k^j) = \Pr(X_1 \ge y_{1k}^j,\ldots,X_n \ge y_{nk}^j) = \prod_{i=1}^{n} \Pr(X_i \ge y_{ik}^j)$ and

$$\Pr(B_l^j) = \Pr(X_1 \le z_{1l}^j,\ldots,X_n \le z_{nl}^j) = \prod_{i=1}^{n} \Pr(X_i \le z_{il}^j).$$

The computational complexity to arrive at the exact system reliability grows exponentially with the number of components. Due to this difficulty, many researchers aim at searching for bounds of the system reliability [5]. A comprehensive summary of the theoretical studies on classical MSMs is presented in [12].

3.2 Generalized Multi-state Models

After the establishment of classical MSM, subsequent efforts have been made to extend its modelling. In 1982 and 1983, Hudson and Kapur [13,14] defined an MSM allowing different numbers of states for each component and the system: $x_i \in S_i = \{0,\ldots,M_i\}$ and $\phi \in S$. In 1993, Aven [15] gave a more general definition, which allowed the states being any non-negative real number: $x_i \in \{x_{i0},\ldots,x_{iM_i}\}$, $x_{i0} = 0$ and $x_{ij} < x_{ik}$ for all $j < k$ and $\phi \in \{\phi_0,\ldots,\phi_M\}$, and $\phi_0 = 0$ and $\phi_j < \phi_k$ for all $j < k$. This definition is closer to reality than the calssical MSM in section 3.1, because the states reflect their values to the customer [16]. Based on this MSM, the definitions of MPVs, MCVs, and performance measures (including reliability) are also presented in [15].

In a series of studies [17–19] from 1996 to 1998, Lisnianski and Levitin have defined a MSM abstracted from the reliability models of power system components [20]. Their modelling is close to Aven's MSM. In this definition, one multi-state component i is characterized by the performance level (or rate), $g_i \in L_i = \{g_{i0},\ldots,g_{iM_i}\}$ where g_{ij} is a non-negative real number and g_{i0}, and g_{iM_i} are the performance levels at complete failure and perfect functioning states, respectively. The elements in the performance set are assumed to be in ascending order. Let the vector $\mathbf{g} = (g_1,\ldots,g_n)$ denote the performances of all components. The system performance $\varphi \in \mathbb{R}$ is a deterministic function of g, i.e. $\varphi = \varphi(\mathbf{g})$. The function $\varphi(\cdot)$ is also called the system structure function and it has the following assumptions:

i φ is non-decreasing in each argument;
ii $\arg\max_g \varphi(g) = \{g_{1M_1},\ldots,g_{nM_n}\}$ and $\arg\min_g \varphi(\mathbf{g}) = \{g_{10},\ldots,g_{n0}\}$; and
iii for all $i \in \{1,\ldots,n\}$ and all $g_{ij} \in L_i$, there exists a vector $(\cdot,\mathbf{g}) = (g_1,\ldots,g_{i-1},\cdot,g_{i+1},\ldots,g_n)$ such that for $j \neq k$, we have $\phi(g_{ij},\mathbf{g}) \neq \phi(g_{ik},\mathbf{g})$.That is, the system contains no irrelevant components.

The properties above are analogous to the assumptions *i*, *ii*, and *iv* of the classical MSS model. Under a random setting, the performance G_i of component *i* is a random variable taking values from L_i, and the system demand W is a random variable taking values from $D = \{w_0,\ldots,w_M \mid w_j \in \mathbb{R}_{\geq 0}\}$, a set of non-negative real-valued demand levels. Then, $p_{ij} = \Pr(G_i = g_{ij})$ is the probability of component *i* being at state *j* and $\sum_{j=0}^{M_i} p_{ij} = 1$ for any *i*. For the system demand, $q_j = \Pr(W = w_j)$ and $\sum_{j=0}^{M} q_j = 1$. Let random vector $G = (G_1,\ldots,G_n)$ denote the state of all components; then, $\varphi = \varphi(G)$ is a random variable representing the system performance. The system reliability with respect to W is defined as

$$R_{\varphi}^{W} = \Pr\left(\varphi \geq W\right) = \sum_{j=0}^{M} q_j \Pr(\varphi \geq w_j \mid W = w_j). \tag{3.2}$$

Lisnianski and Levitin's MSM have become the most frequently applied and studied in recent literature because they give a representation closer to the reality compared to previous MSMs. More details about this modelare presented in their book [21].

3.3 Time-dependent Multi-state Models

In practice, multi-state components or MSSs may be requested to work at different performance levels at different times to satisfy the customer's changing demand. In addition, they may be subject to maintenance actions, which require them to be off-line for a period of time. For examples, the power plant production varies according to the daily or seasonal load demands, and the plant needs to be regularly shut down for inspections and repairs. This stimulates the efforts to consider the time dimension in MSMs and to develop different reliability measures for the dynamic MSMs.

In 1984, Funnemark and Natvig [22] considered the time-dependent MSM (TDMSM). The state of component i is a stochastic process $X_i(t)$. At any fixed time $t \in \mathbb{R}_{\geq 0}$, $X_i(t)$ is a random variable taking values from S_i. The joint state of all components $\boldsymbol{X}(t) = (X_1(t),\ldots,X_n(t))$ follows a vector stochastic process. The state of an MSS with the structure function $\phi$is a stochastic process $\phi(\boldsymbol{X}(t))$. At any fixed time t, $\phi(\boldsymbol{X}(t))$ is a random variable taking values from S. The MPVs and MCVs are defined at the fixed time t, similar to those in the time-independent MSMs (TIMSMs). The generalized TDMSMs [15,21] basically reuse the definitions made by Funnemark and Natvig. The only difference of Aven's TDMSM [15] is that the elements of the sets where $X_i(t)$ and $\phi(X(t))$aretaking values from can be non-negative real numbers. In Lisnianski and Levitin's TDMSM [21], $G_i(t)$, $\varphi(\boldsymbol{G}(t))$and $W(t)$ are used to denote the stochastic component performance, system performance, and system demand, respectively. They all can take non-negative real values.

In reliability engineering, various types of stochastic processes, e.g. point processes, renewal processes, and Markov processes [21,23,24] are applied to model component dynamics among which Markov processes are mostly used [21]. In the family of Markov Processes, continuous time Markov chain (CTMC) is the most applied one. In the case of discrete time, e.g. $t_n = t_{n-1} + \Delta t$, the discrete time Markov chain (DTMC) is used. For simpilicity, we will only discuss the about the continuous case, in this section. Let $G_i(t)$ be a CTMC on the set L_i; the quantity of primary interest is the state probability vector $p_i(t) = (p_{i0}(t),\ldots,p_{iM_i}(t))$ at any time t. By the law of probability, we have $\sum_{j=0}^{M_i} p_{ij}(t) = 1$ at any time t. In the case of homogeneous CTMC (HCTMC), $p_i(t)$ can be found by solving the following system of differential equations:

$$\frac{d}{dt}p_j^i(t)=\sum_{\substack{k=0\\k\neq j}}^{M_i}p_k^i(t)\lambda_{kj}^i-p_j^i(t)\sum_{\substack{k=0\\k\neq j}}^{M_i}\lambda_{jk}^i, \tag{3.3}$$

where $p_j^i(t)$, the same as $p_{ij}(t)$, is the probability of state j of component i at time t (for ease of notation, we move the index of the component to the superscript), and λ_{kj}^i is the rate which characterizes the stochastic transition of component i from state k to state j. The transition rate λ_{kj}^i is defined as

$$\lambda_{kj}^i=\lim_{\Delta t\to 0}\frac{\Pr(G_i(t+\Delta t)=g_{ij}\mid G_i(t)=g_{ik})}{\Delta t}. \tag{3.4}$$

HCTMC is applicable only when the transition time between any two states, i.e. the state holding time, follows an exponential distribution. In many real-world cases, this restriction needs to be removed. Semi-Markov process (SMP) is an alternative because it allows arbitrary state holding time distributions [21]. The key concept of SMP is the kernel $Q_{kj}^i(t)$, analogous to λ_{kj}^i of the CTMC:

$$\begin{aligned}Q_{kj}^i(t)&=\Pr\left(G_{n+1}^i=g_{ij},T_{n+1}^i-T_n^i\leq t\mid G_0^i,\ldots,G_n^i;T_0^i,\ldots,T_n^i\right)\\&=\Pr\left(G_{n+1}^i=g_{ij},T_{n+1}^i-T_n^i\leq t\mid G_n^i=g_{ik}\right),\end{aligned} \tag{3.5}$$

where T_n^i is the time of the n-th transition of component i, and G_n^i is the performance of component i at the n-th transition. Similar to CTMC, the SMP is mainly used to find the component state probabilities. Let $\theta_{jk}^i(t)$ denote the probability that the process of component i starts from state j at time 0 and will reach state k at time t. By solving the following system of integral equations, it can be found that

$$\theta_{jk}^i(t)=\delta_{jk}^i\left[1-F_j^i(t)\right]+\sum_{m=0}^{M_i}\int_0^t q_{jm}^i(\tau)\theta_{mk}^i(t-\tau)d\tau, \tag{3.6}$$

where $\delta_{jk}^i=1$ if $j=k$ and $\delta_{jk}^i=0$ if $j\neq k$, $F_j^i(t)=\sum_{m=0}^{M_i}Q_{jm}^i(t)$, and $q_{jm}^i(\tau)=\frac{dQ_{jm}^i(\tau)}{d\tau}$.

Given the initial state and the kernel matrix $\left|Q_{kj}^i(t)\right|$, $p_j^i(t)$ can be found. The CTMC and SMP can also be used to solve for the system state probability when the transition rates and kernel matrixes for the system transitions are defined.

As to the system reliability-related measure, availability is the most frequently used metric in TDMSMs. It quantifies the ability of the system to satisfy the customer demand at any specific moment during the system life time. In generalized TDMSMs [15,21], the availability is defined assuming a minimum on total performance of φ as

$$A_\varphi^W(t)=\sum_{j=0}^{M}\Pr(\varphi(G(t))\geq w_j\mid W(t)=w_j)\times\Pr(W(t)=w_j). \tag{3.7}$$

By extending the time-dependent instantaneous availability in Equation (3.7), other time-independent measures, such as average and limiting availability [15,25], can be introduced to quantify the integral and asymptotic measure of the system reliability characteristics (i.e., the probability that the system is in the desired state or above during a time interval or after the initial transient, asymptotically in time).

3.4 Methods to Evaluate Multi-state System Reliability

Based on MSMs, a number of reliability assessment methods have been proposed. In this section, they are classified into four groups. They are mainly developed for the TIMSMs. Some of them, e.g. the methods based on MPVs or MCVs and the universal generating function (UGF) approach are extended to TDMSMs by Natvig [12] and Lisnianski and Levitin [21], respectively. The Monte Carlo simulation (MCS) method is naturally adapted to the time-dependent case as proposed by Zio, et al. [26].

3.4.1 Methods Based on MPVs or MCVs

As mentioned in Section 3.1, in classical multi-state theory, the system reliability is expressed in terms of the probability of the union of all MPVs or all MCVs. Based on this formulation, two ways exist to evaluate system reliability. The first is to mathematically derive the lower and upper bounds [5,8,9,27]. The second computes the exact reliability using special principles or algorithms, such as inclusion-exclusion method [5,10,14,28], state-space decomposition [29], and recursive method [30]. A common prerequisite of these methods is that all MPVs or MCVs of an MSS are given. However, finding all of them is, in general, computationally difficult despite some algorithms are proposed for special classes of MSS [31,32].

3.4.2 Methods Derived from Binary State Reliability Assessment

Representing the multi-state component by a set of binary variables, the MSS reliability can be eventually calculated using the well-established binary algorithms. In 1980, Caldarola [33] proposed the Boolean algebra with restrictions on variables for this conversion. The binary variable b_{ij} takes the value of 1 if component i is at state j and the value of 0 if it is at one state other than j. There are two restrictions on each b_{ij}: 1) $\vee_{j=0}^{M_i} b_{ij} = 1$ and 2) $b_{ij} \wedge b_{ik} = 0$, $\forall j \neq k$. With these restrictions, the basic rules of traditional Boolean algebra operations are applied to derive the system state expression. b_{ij}s are not pairwise mutually independent and the number of binary variables is $(m_1+1)\times\ldots\times(m_n+1)$ for representing the system state. In [34], this method is adopted to analyze MSS, and the inclusion-exclusion method is, then, used to obtain system probability expressions.

In 1994, Wood [35] proposed a slightly different conversion. The state of component i is represented by the sum of b_{ij} and i.e. $x_i = \sum_{j=0}^{M_i} b_{ij}$, and the state of the system is represented by the sum of binary variables $\phi^j(b)$ and i.e. $\phi(b)=\sum_{j=0}^{M}\phi^j(b)$. The restrictions are that $b_{ij}=1$ implies $b_{ik}=1$ for all $k \le j$ and $\phi^j(b)=1$implies $\phi^k(b)=1$ for all $k \le j$. This conversion has the same drawbacks as Caldarola's approach. To compute system reliability, the conditional probability expansion (i.e. factoring approach) is used in [35] to handle the following situations: the components are dependent and the same variable appears in multiple places in the system expression.

Fault trees are often used to find the state probability distributions of binary systems. In 1990, Kai [36] applied the recursive pivotal decomposition algorithm of binary fault trees to the multi-state case. This method does not require the MPVs or MCVs.

Binary decision diagrams (BDDs), proposed by Bryant [37], are an efficient tool of Boolean expressions manipulation because they require less memory to represent large Boolean expressions compared to other methods. In 2003, Zhang, et al. [38] applied BDDs for MSS reliability assessment. The basis for BDD implementation is the Shannon decomposition:

Let f be a Boolean logic function on the set of binary variables $\{b_1,\ldots,b_m\}$. Then

$$f=(b_i=0)f_{b_i=0}+(b_i=1)f_{b_i=1} \quad \text{for any} \quad i, \tag{3.8}$$

where $f_{b_i=v}$ is f evaluated with $b_i=v$. Its idea is similar to the factoring approach. In [38], a BDD operation is proposed to realize the Boolean algebra with restrictions on variables, and the final system BDD is efficiently evaluated to obtain the system reliability. However, this approach still involves a large number of possibly dependent binary variables. To remedy this problem, the multi-valued decision diagram (MDD) [39] is applied. This approach directly uses the multi-valued x_i and implements a multi-valued logic function. In a more recent work [40], MDD is used to evaluate multi-state k-out-of-n system reliability in comparison with the recursive method proposed in [30].

3.4.3 Universal Generating Function Approach

The UGF approach, originated by Ushakov in 1987 [41], is adopted for system reliability assessment by Lisnianski, Levitin, and their colleagues in [17], based on their MSS model. It is an analytical tool to describe multi-state components and to construct the overall model of complex MSSs. The UGF of component i is expressed as

$$u_i(z)=\sum_{j=0}^{M_i}p_{ij}z^{g_{ij}}, \tag{3.9}$$

where z is the base of the z-transform. It is essentially an equivalent representation of the probability mass function (pmf) of the performance of component i. Based on the component UGF, the composition operator $\otimes_\varphi$ is proposed to derive the UGF of an arbitrary MSS with the structure function $\varphi(G_1, G_2, \ldots G_n)$ [42]. It has the following general expression

$$\otimes_{\varphi}\left(u_1(z), \ldots, u_n(z)\right) = \sum_{j_1=0}^{M_1} \cdots \sum_{j_n=0}^{M_n} \left(\prod_{j_i=0}^{M_i} p_{ij_i} z^{\varphi(g_{1j_1}, \ldots, g_{nj_n})} \right). \tag{3.10}$$

To derive the system UGF, the iterative approach [42] is often used:

$$u_{2'}(z) = \otimes_{\varphi_{1\&2}}\left(u_1(z), u_2(z)\right),\ u_{3'}(z) = \otimes_{\varphi_{2'\&3}}\left(u_{2'}(z), u_3(z)\right), \ldots,$$

$$u_{n'}(z) = u_{\varphi}(z) = \otimes_{\varphi_{(n-1)'\&n}}\left(u_{(n-1)'}(z), u_n(z)\right)$$

where 2', 3', ..., n' are the virtual components, essentially the combinations of 2, 3, ..., n components, respectively. The like-term collection technique [43] is implemented during each iteration to enhance computation efficiency. The sequence of components in the iterations also affects computational speed.

In 2008, Li and Zuo [44] proposed a recursive algorithm with the following formula:

$$R_k(w_k, n) = \sum_{j=0}^{M_n} p_{nj} R_k(w_k - g_{nj}, n-1), \tag{3.11}$$

where k denotes the k-th system demand state to compete with the UGF composition approach for the reliability evaluation of multi-state weighted k-out-of-n systems. The results show that the recursive algorithm is generally more efficient than UGF composition though the time complexities of the two approaches are both exponential to n in the worst cases.

3.4.4 Monte Carlo Simulation

The application of the above evaluation methods generally have certain prerequisites, e.g. independence of components. However, many real-world systems often possess complex characteristics, e.g. operational dependencies. For example, in a production line of a nodal series structure, if one of the nodes throughput changes (e.g. switches from 100% to 50%), the other nodes will have to be reconfigured (i.e. they must deterministically change their states) so as to provide the same throughput [45]. The MCS appears to be the only feasible approach to quantitatively capture the realistic aspects of the MSS complex dynamic behavior [46]. In 2003, Zio et al. [45] proposed an MCS technique which allows modeling multi-state components subject to operational dependencies. Later, MCS is used to estimate the reliability of a multi-state network by Ramirez-Marquez and Coit [47]. The major disadvantage of MCS is the uncertainty in its convergence to a stable estimate within reasonable computational time. The successful implementation of MCS depends on the proper representation and modeling of the multi-state dynamics of the components and the systems, e.g. by Petri Nets [48], and on the efficient evaluation methods, e.g. by biasing techniques [49].

3.5 Exercises

1 Find all minimal path and cut sets and compute the reliability of the following system: number of components, $N=3$; highest state, $M=2$; $\mathrm{P}(x_i \geq 1)=0.9$, $\mathrm{P}(x_i=2)=0.7$, for $i=1, 2, 3$; $\phi(x)=\min(x_1,(x_2+x_3))$; $R=\Pr(\phi(x)\geq 1)$.

2 Derive the following integral equation for semi-Markov state probability

$$\theta_{jk}^{i}(t)=\delta_{jk}^{i}\left[1-F_j^i(t)\right]+\sum_{m=0}^{M_i}\int_0^t q_{jm}^{i}(\tau)\theta_{mk}^{i}(t-\tau)d\tau.$$

3 Consider a multi-state parallel system with three components. Every component has three possible states: 0, 1, 2. The system function is $\phi(g)=g_1+g_2+g_3$. The component state performance and probability distributions are shown in the following tables. Compute the system reliability $R=\Pr(\phi(x)\geq 5)$ using MCS and UGF methods.

State probability distribution of each component

	$j=0$	$j=1$	$j=2$
$i=1$	0.1	0.2	0.7
$i=2$	0.4	0.2	0.4
$i=3$	0.3	0.5	0.2

State performance of each component

	$j=0$	$j=1$	$j=2$
$i=1$	1	2	3
$i=2$	1	3	4
$i=3$	1	3	5

References

1 Barlow, R.E. and Proschan, F. (1975). *Statistical Theory of Reliability and Life Testing: Probability Models*. New York City: Holt, Rinehart and Winston.

2 Natvig, B. and Mørch, H.W. (2003). An application of multistate reliability theory to an offshore gas pipeline network. *International Journal of Reliability, Quality and Safety Engineering* 10 (04): 361–381.

3 Li, Y.F., Valla, S., and Zio, E. (2015). Reliability assessment of generic geared wind turbines by GTST-MLD model and Monte Carlo simulation. *Renewable Energy* 83: 222–233.

4 Barlow, R.E. and Wu, A.S. (1978). Coherent systems with multi-state components. *Mathematics of Operations Research* 3 (4): 275–281.

5 El-Neweihi, E., Proschan, F., and Sethuraman, J. (1978). Multistate coherent systems. *Journal of Applied Probability* 15 (4): 675–688.

6 Ross, S.M. (1979). Multi-valued state component reliability systems. *Annals of Probability* 7: 379–383.

7 Griffith, W.S. (1980). Multistate reliability models. *Journal of Applied Probability* 17 (3): 735–744.

8 Block, H.W. and Savits, T.H. (1982). A decomposition for multistate monotone systems. *Journal of Applied Probability* 19 (2): 391–402.

9 Butler, D.A. (1982). Bounding the reliability of multistate systems. *Operations Research* 30 (3): 530–544.

10 Natvig, B. (1982). Two suggestions of how to define a multistate coherent system. *Advances in Applied Probability* 14 (2): 434–455.

11 Ebrahimi, N. (1984). Multistate reliability models. *Naval Research Logistics Quarterly* 31 (4): 671–680.

12 Natvig, B. (2010). *Multistate Systems Reliability Theory with Applications*. John Wiley & Sons.

13 Hudson, J.C. and Kapur, K.C. (1982). Reliability theory for multistate systems with multistate components. *Microelectronics Reliability* 22 (1): 1–7.

14 Hudson, J.C. and Kapur, K.C. (1983). Reliability analysis for multistate systems with multistate components. *AIIE Transactions* 15 (2): 127–135.

15 Aven, T. (1993). On performance measures for multistate monotone systems. *Reliability Engineering & System Safety* 41 (3): 259–266.

16 Brunelle, R.D. and Kapur, K.C. (1997). Customer-centered reliability methodology. In: *Reliability and Maintainability Symposium. 1997 Proceedings, Annual*, 286–292. IEEE.

17 Lisnianski, A., Levitin, G., Ben-Haim, H., and Elmakis, D. (1996). Power system structure optimization subject to reliability constraints. *Electric Power Systems Research* 39 (2): 145–152.

18 Levitin, G., Lisnianski, A., and Elmakis, D. (1997). Structure optimization of power system with different redundant elements. *Electric Power Systems Research* 43 (1): 19–27.

19 Levitin, G., Lisnianski, A., Ben-Haim, H., and Elmakis, D. (1998). Redundancy optimization for series-parallel multi-state systems. *IEEE Transactions on Reliability* 47 (2): 165–172.

20 Billinton, R., Allan, R.N., and Allan, R.N. (1984). *Reliability Evaluation of Power Systems*. New York City: Plenum press.

21 Lisnianski, A. and Levitin, G. (2003). *Multi-state System Reliability: Assessment, Optimization and Applications*, 6. World scientific.

22 Natvig, B. and Streller, A. (1984). The steady-state behaviour of multistate monotone systems. *Journal of Applied Probability* 21 (4): 826–835.

23 Aven, T. and Jensen, U. (1999). *Stochastic Models in Reliability*. New York City: Springer.

24 Epstein, B. and Weissman, I. (2010). *Mathematical Models for Systems Reliability*. CRC Press.

25 Brunelle, R.D. and Kapur, K.C. (1999). Review and classification of reliability measures for multistate and continuum models. *IIE Transactions* 31 (12): 1171–1180.

26 Zio, E. and Podofillini, L. (2003). Monte Carlo simulation analysis of the effects of different system performance levels on the importance of multi-state components. *Reliability Engineering & System Safety* 82 (1): 63–73.

27 Hudson, J.C. and Kapur, K.C. (1985). Reliability bounds for multistate systems with multistate components. *Operations Research* 33 (1): 153–160.

28 Janan, X. (1985). On multistate system analysis. *IEEE Transactions on Reliability* 34 (4): 329–337.

29 Aven, T. (1985). Reliability evaluation of multi-state systems with multi-state components. *IEEE Transactions on Reliability* 34 (5): 473–479.

30 Zuo, M.J. and Tian, Z. (2006). Performance evaluation of generalized multi-state k-out-of-n systems. *IEEE Transactions on Reliability* 55 (2): 319–327.

31 Yeh, W.-C. (2001). A simple approach to search for all *d*-MCs of a limited-flow network. *Reliability Engineering & System Safety* 71 (1): 15–19.

32 Yeh, W.-C. (2001). A simple algorithm to search for all *d*-MPs with unreliable nodes. *Reliability Engineering & System Safety* 73 (1): 49–54.

33 Caldarola, L. (1980). Coherent systems with multistate components. *Nuclear Engineering and Design* 58 (1): 127–139.

34 Veeraraghavan, M. and Trivedi, K.S. (1994). A combinatorial algorithm for performance and reliability analysis using multistate models. *IEEE Transactions on Computers* 43 (2): 229–234.

35 Wood, A.P. (1985). Multistate block diagrams and fault trees. *IEEE Transactions on Reliability* 34 (3): 236–240.

36 Kai, Y. (1990). Multistate fault-tree analysis. *Reliability Engineering & System Safety* 28 (1): 1–7.

37 Bryant, R.E. (1986). Graph-based algorithms for boolean function manipulation. *IEEE Transactions on Computers* 100 (8): 677–691.

38 Zang, X., Wang, D., Sun, H., and Trivedi, K.S. (2003). A BDD-based algorithm for analysis of multistate systems with multistate components. *IEEE Transactions on Computers* 52 (12): 1608–1618.

39 Xing, L. and Dai, Y. (2009). A new decision-diagram-based method for efficient analysis on multistate systems. *IEEE Transactions on Dependable and Secure Computing* 6 (3): 161–174.

40 Mo, Y.C., Xing, L.D., Amari, S.V., and Dugan, J.B. (2015). Efficient analysis of multi-state k-out-of-n systems. *Reliability Engineering & System Safety* 133: 95–105.

41 Ushakov, I. (1987). Optimal standby problems and a universal generating function. *Soviet Journal of Computer and Systems Sciences* 25 (4): 79–82.

42 Levitin, G. (2006). *The Universal Generating Function in Reliability Analysis and Optimization*. London: Springer.

43 Levitin, G. (2001). Reliability evaluation for acyclic consecutively connected networks with multistate elements. *Reliability Engineering & System Safety* 73 (2): 137–143.

44 Li, W. and Zuo, M.J. (2008). Reliability evaluation of multi-state weighted k-out-of-n systems. *Reliability Engineering & System Safety* 93 (1): 160–167.

45 Zio, E., Marella, M., and Podofillini, L. (2007). A Monte Carlo simulation approach to the availability assessment of multi-state systems with operational dependencies. *Reliability Engineering & System Safety* 92 (7): 871–882.

46 Poszgai, P. and Bertsche, B. (2003). On the influence of the passive states on the availability of mechanical systems. In: *Safety and Reliability* (ed. Bedford & van Gelder), 1255–1262.

47 Ramirez-Marquez, J.E. and Coit, D.W. (2005). A Monte-Carlo simulation approach for approximating multi-state two-terminal reliability. *Reliability Engineering & System Safety* 87 (2): 253–264.

48 Dutuit, Y., Châtelet, E., Signoret, J.-P., and Thomas, P. (1997). Dependability modelling and evaluation by using stochastic Petri nets: Application to two test cases. *Reliability Engineering & System Safety* 55 (2): 117–124.

49 Marseguerra, M. and Zio, E. (1993). Nonlinear Monte Carlo reliability analysis with biasing towards top event. *Reliability Engineering & System Safety* 40 (1): 31–42.

4

Markov Processes

As discussed in Chapter 3, a multi-state system (MSM) is often applied for system degradation process modeling because it offers the possibility of describing the degradation state of the components and system by a number of consecutive levels from perfect working to complete failure. To model the dynamics and transitions of such a multi-state degradation process, Markov models have often been used [1-3]. In doing this, the rates of transition among the degradation states are typically assumed to be constant, which implies that the degradation process is memoryless. The resulting stochastic process is called a homogeneous continuous time Markov chain (HCTMC). In many realistic situations, (e.g. cracking of nuclear component [4], battery aging [5], and cancer patients' life quality [6]) with varying external factors influencing the degradation processes, the transition rates can no longer be considered as time-independent. Under these circumstances, the inhomogeneous continuous time Markov chain (ICTMC) is more suited than HCTMC for modeling the degradation process. In addition, the semi-Markov process (SMP) is a more general model than the ICTMC that can deal with arbitrary transition rates: given this property, it is well-suited to model the degradation influenced by the environmental factors that change with time and other features. Piecewise-deterministic Markov process (PDMP) can be regarded as a special class of the SMP that explicitly describe the system dynamics and the degradation dependence.

In all, the theoretical foundations of Markov processes presented in this chapter are kept to the minimum, meaning brief and limited. Interested readers can refer to any textbook on stochastic processes for more details.

4.1 Continuous Time Markov Chain (CMTC)

Markov chain is a special type of Markov process. Let $\{X(t), t \geq 0\}$ denote a stochastic process defined on a state space $S = \{0, 1, \ldots, M\}$, which is finite or infinite. Assume the state of the process at time t is $X(t) = i$; the conditional probability that the process will be in state j at time $t + \Delta t$ is

Reliability Analysis, Safety Assessment and Optimization: Methods and Applications in Energy Systems and Other Applications, First Edition. Enrico Zio and Yan-Fu Li.

$$\Pr\left(X\left(t+\Delta t\right)=j \mid X\left(t\right)=i, X\left(u\right)=h\left(u\right), 0 \le u < t\right), \tag{4.1}$$

where $h(u)$ is the historical trajectory of the process till time t. In real-life situations, keeping the complete history of the process is often difficult; thus, it is reasonable to assume that the future evolution of the process is only dependent on the present situation and independent of anything that has happened in the past. In mathematical words, we have the following equation

$$\Pr\left(X\left(t+\Delta t\right)=j \mid X\left(t\right)=i, X\left(u\right)=h\left(u\right), 0 \le u < t\right)=\Pr\left(X\left(t+\Delta t\right)=j \mid X\left(t\right)=i\right). \tag{4.2}$$

This is called the Markov property. A stochastic process that satisfies the Markov property, i.e. Equation (4.2), is called a Markov process (or a CTMC). When the time t is discrete, it is called a discrete time Markov chain (DTMC). In DTMC, the time is often denoted by a step indicator k and the chain by $\{X_k, k=0,1,\ldots,L\}$ where L is the index of the last time step. Because CTMC is more frequently used in reliability engineering and DTMC has several similarities to CTMC, we will focus on CTMC in this book.

The conditional probability $\Pr\left(X\left(t+\Delta t\right)=j \mid X\left(t\right)=i\right)$ in Equation (4.2) represents the probability that the process will, when in state i at time t, make a transition into state j at time $t+\Delta t$. This value is called a one-step transition probability, denoted as $p_{ij}\left(t,t+\Delta t\right)$. If $p_{ij}\left(t,t+\Delta t\right)$ is independent of t, then the CTMC is said to have stationary or homogeneous transition probabilities. In a mathematical expression, we have the following equation:

$$\Pr\left(X\left(t+\Delta t\right)=j \mid X\left(t\right)=i\right)=\Pr\left(X\left(\Delta t\right)=j \mid X\left(0\right)=i\right), \forall t, \Delta t \ge 0. \tag{4.3}$$

It is called the stationary property of the Markov process. CTMC with the stationary property is called HCTMC. The transition probability $p_{ij}\left(s,s+t\right)$ in HCTMC is briefly denoted as $p_{ij}\left(t\right)$. The matrix $\boldsymbol{P}\left(t\right)=\left(p_{ij}\left(t\right), i,j \in S, t \ge 0\right)$ is called the transition probability matrix of the HCTMC.

For $p_{ij}\left(t\right)$ of an HCTMC, the following limits exist:

$$\lim_{\Delta t \to 0} \frac{1-p_{ii}\left(\Delta t\right)}{\Delta t}=v_i=\lambda_{ii}, \tag{4.4}$$

$$\lim_{\Delta t \to 0} \frac{p_{ij}\left(\Delta t\right)}{\Delta t}=\lambda_{ij}, i \ne j, \tag{4.5}$$

where λ_{ij} is called the transition rate from state i to state j, and v_i is called the transition rate associated with state i. Let τ_i denote the sojourn time of X in state i before making a transition to a different state; τ_i follows the exponential distribution with parameter v_i. State i is named as absorbing if $v_i=0$, named as stable if $0 < v_i < \infty$, and instantaneous if $v_i=\infty$.

The transition rates of the HCTMC form a matrix $\boldsymbol{\Lambda}=\left(\lambda_{ij}\right)$:

$$= \begin{bmatrix} -\lambda_{00} & \lambda_{01} & \cdots & \lambda_{0M} \\ \lambda_{10} & -\lambda_{11} & \cdots & \lambda_{1M} \\ \vdots & \vdots & \ddots & \vdots \\ \lambda_{M0} & \lambda_{M1} & \cdots & -\lambda_{MM} \end{bmatrix}. \tag{4.6}$$

Clearly, we have $\boldsymbol{\Lambda I} = 0$ where $\boldsymbol{I}$ denotes the identity matrix, which indicates that the row-sums of $\boldsymbol{\Lambda}$ are equal to zero. From Equations (4.4) and (4.5), it yields that $\boldsymbol{\Lambda} = \boldsymbol{P}'(0)$. The transition rate matrix $\boldsymbol{\Lambda}$ and the initial state probability vector $\mathbf{p}(0) = (p_i(0), i \in S)$, then, completely characterize a CTMC.

Example 4.1 [7]

Consider a system with one component and one repairman. Assume the failure times and repair times of the system follow exponential distributions. The component has a discrete state space $S = \{0,1\}$ where state 0 represents the working state and state 1 represents the failure state. The rate of failure (i.e. the transition rate from state 0 to state 1) of the system is λ, and the rate of repair (i.e. the transition rate from state 1 to state 0) of the system is μ. The Markov diagram of this system is sketched in Figure 4.1.

Because we have $\boldsymbol{\Lambda I} = 0$, the transition rate matrix of the one component/one repairman system is given as follows:

$$\boldsymbol{\Lambda} = \begin{bmatrix} -\lambda & \lambda \\ \mu & -\mu \end{bmatrix}.$$

The primary quantity of interest in many applications of CTMC is the state probability vector at any time instant t, $\mathbf{p}(t) = (p_i(t), i \in S, t \geq 0)$. By the definition of probability, we have $\sum_{i \in S} p_i(t) = 1$, $\forall t \geq 0$. Obviously, $p_i(t) = \sum_{j \in S} p_j(0) p_{ji}(t)$. In the case of HCTMC, $\mathbf{p}(t)$ is typically found by solving the following system of differential equations (i.e. Chapman-Kolmogorov equations)

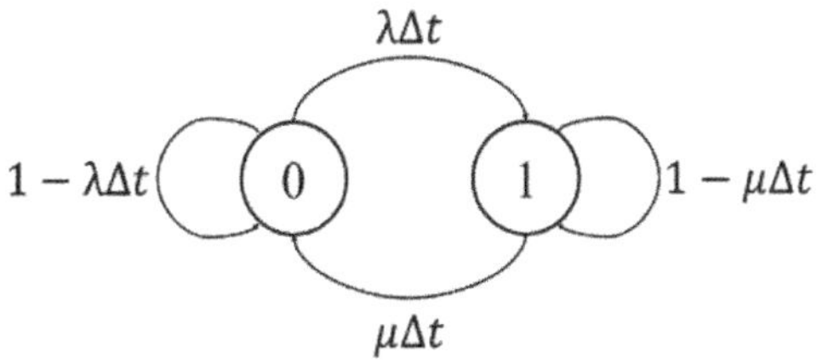

Figure 4.1 The Markov Diagram of the One Component/one Repairman System.

$$\frac{d}{dt}p_i(t)=\sum_{\substack{j\in S\\ j\neq i}}p_j(t)\lambda_{ji}-p_i(t)\sum_{\substack{j\in S\\ j\neq i}}\lambda_{ij}, \tag{4.7}$$

where i,j are the state indexes ranging from 0 to M, and λ_{ij} is the transition rate of HCTMC.

To obtain Equation (4.7), we first decompose the state probability as

$$p_i(t+\Delta t)=\sum_{\substack{j\in S\\ j\neq i}}p_{ji}(\Delta t)p_j(t)+p_{ii}(\Delta t)p_i(t). \tag{4.8}$$

Subtracting $p_i(t)$ from both sides yields

$$p_i(t+\Delta t)-p_i(t)=\sum_{\substack{j\in S\\ j\neq i}}p_{ji}(\Delta t)p_j(t)-(1-p_{ii}(\Delta t))p_i(t). \tag{4.9}$$

Dividing Equation (4.9) by Δt and letting $\Delta t\to 0$ yields Equation (4.7).

Example 4.2

Reference [7] considers a system with M identical components and M repairmen available. Assume that each component can be in two states: working or failure. The rate of failure and rate of repair are λ and μ, respectively. The state space of the M components and M repairmen system is $S=\{i\in\mathbb{N}\mid 0\le i\le M\}$ where $\mathbb{N}$ denotes the set of non-negative integer numbers. State i shows there are i components are currently failed in the system and the others are functioning. Thus, state 0 represents the case that all components are functioning and state M represents that all components have failed. Besides, we propose the following hypotheses: No more than one event (i.e. the failure or repair of a component) can simultaneously occur in a sufficiently small time interval Δt, such that all events are mutually exclusive.

From the Chapman-Kolmogorov equations [8], it is clear that the transition rate matrix could be constructed as long as one derives the transition probability between state i and its successive state $i+1$. Let us consider the transition probability in the sufficiently-small time interval Δt,

$$\begin{aligned}
p_{i,i+1}(\Delta t) &= \Pr(\text{anyone of the } M-i \text{ components fails in } \Delta t \mid \\
&\qquad \text{there are already } i \text{ components failed}) \\
&= \Pr(\text{the 1st working component fails in } \Delta t) \\
&\quad +\Pr(\text{the 2nd working component fails in } \Delta t)+\ldots \\
&\quad +\Pr(\text{the } (M-i)\text{th working component fails in } \Delta t) \\
&= \lambda\Delta t+\lambda\Delta t+\ldots+\lambda\Delta t \\
&= (M-i)\lambda\Delta t
\end{aligned}$$

Similarly,

$$p_{i+1,i}(\Delta t) = (i+1)\mu\Delta t$$

Hence, one gets the following transition rates:

$$\lambda_{i,i+1} = (M-i)\lambda$$

$$\lambda_{i,i-1} = i\mu$$

$$\lambda_{ii} = \lambda_{i,i+1} + \lambda_{i,i-1} = (M-i)\lambda + i\mu$$

The transition rate matrix of the M components and M repairmen system is

$$\boldsymbol{\Lambda} = \begin{bmatrix} -M\lambda & M\lambda & & & & \\ \mu & -\mu-(M-1)\lambda & (M-1)\lambda & & & \\ & 2\mu & -2\mu-(M-2)\lambda & (M-2)\lambda & & \\ & & \ddots & \ddots & \ddots & \\ & & & (M-1)\mu & -(M-1)\mu-\lambda & \lambda \\ & & & & M\mu & -M\mu \end{bmatrix}$$

The Markov diagram of the M components and M repairmen system is sketched in Figure 4.2.

For the finite state space, Equation (4.7) can be solved by several different approaches. In this chapter, we introduce the classical Laplace transform method [9]. Let us define the Laplace transform of the state probability $p_i(t)$ as

$$p_i^e(s) = \int_0^\infty e^{-st} p_i(t)\,dt \tag{4.10}$$

and the matrix function $\mathbf{p}^e(s) = (p_i^e(s), i \in S)$. We take the Laplace transform of Equation (4.7) and derive

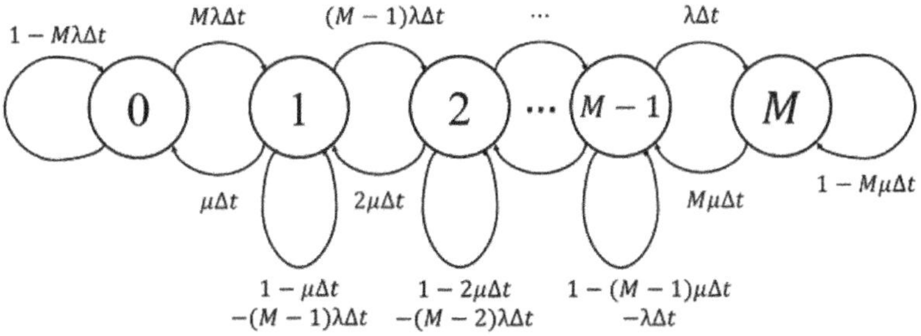

Figure 4.2 The Markov Diagram of the M Components and M Repairmen System.

$$s\boldsymbol{p}^e(s) - \boldsymbol{p}(0) = \boldsymbol{\Lambda}\boldsymbol{p}^e(s), \tag{4.11a}$$

$$[s\boldsymbol{I} - \boldsymbol{\Lambda}]\boldsymbol{p}^e(s) = \boldsymbol{p}(0). \tag{4.11b}$$

Hence, $\boldsymbol{p}^e(s) = \boldsymbol{p}(0)[s\boldsymbol{I} - \boldsymbol{\Lambda}]^{-1}$ where $[\cdot]^{-1}$ denotes the inverse matrix. By inverting the Laplace transform $\boldsymbol{p}^e(s)$ back to the time domain, we could obtain the state probability vector $\boldsymbol{p}(t)$.

When $t \to \infty$, the state probabilities $\boldsymbol{p}_i(t)$ of CTMC may reach steady values πi, which are named steady-state probabilities. To compute the steady-state probabilities, we want the left-hand side of Equation (4.7) to equal zero. Then, the following relation holds

$$\sum_{\substack{j \in S \\ j \neq i}} \pi_j \lambda_{ji} - \pi_i \lambda_{ii} = 0 \tag{4.12}$$

Denote the row vector $\pi = (\pi_i, i \in S)$. From Equation (4.12), we obtain

$$\pi \Lambda = 0, \tag{4.13}$$

$$\pi \boldsymbol{e}^T = 1, \tag{4.14}$$

where $\boldsymbol{e} = (1, \ldots, 1)$ is the unit vector. Equation (4.14) is the normalizing equation of π. By solving Equations (4.13) and (4.14), we obtain

$$\pi = \frac{D_i}{\sum_{j \in S} D_j}, \forall i \in S, \tag{4.15}$$

where D_i represents the determinant of the square matrix obtained from $\boldsymbol{\Lambda}$, by deleting the i-th row and column.

Example 4.3

We consider again the one component and one repairman system. Assume the component is working at $t = 0$, so the initial state probability vector is $\boldsymbol{p}(0) = (1, 0)$. According to Equation (4.11b), we have

$$\begin{aligned}
\boldsymbol{p}^e(s) &= \boldsymbol{p}(0)[s\boldsymbol{I} - \boldsymbol{\Lambda}]^{-1} \\
&= \begin{bmatrix} 1 & 0 \end{bmatrix} \begin{bmatrix} s+\lambda & -\lambda \\ -\mu & s+\mu \end{bmatrix}^{-1} \\
&= \frac{1}{s^2 + s\lambda + s\mu} \begin{bmatrix} 1 & 0 \end{bmatrix} \begin{bmatrix} s+\mu & \lambda \\ \mu & s+\lambda \end{bmatrix} \\
&= \begin{bmatrix} \dfrac{s+\mu}{s(s+\lambda+\mu)} & \dfrac{\lambda}{s(s+\lambda+\mu)} \end{bmatrix}
\end{aligned}$$

We invert the Laplace transform and derive

$$\boldsymbol{p}(t)=\left[\frac{\mu}{\lambda+\mu}+\frac{\lambda}{\lambda+\mu}e^{-(\lambda+\mu)t} \quad \frac{\lambda}{\lambda+\mu}-\frac{\lambda}{\lambda+\mu}e^{-(\lambda+\mu)t}\right]$$

The first vector element is the system instantaneous availability (i.e. the probability of being in operational state 0 at time t), which can be given as

$$p_0(t)=\frac{\mu}{\lambda+\mu}+\frac{\lambda}{\lambda+\mu}e^{-(\lambda+\mu)t}.$$

and the second vector element is the system instantaneous unavailability (i.e. the probability of being in failed state 1 at time t)

$$p_1(t)=\frac{\lambda}{\lambda+\mu}-\frac{\lambda}{\lambda+\mu}e^{-(\lambda+\mu)t}.$$

4.2 In homogeneous Continuous Time Markov Chain

In a HCTMC, the transition rates are constant. However, in many realistic situations (e.g. cracking of nuclear component and battery aging), the external factors influencing the degradation processes are changing, so the transition rates can no longer be regarded as time-independent. Under these circumstances, the ICTMC is more suited than HCTMC for modeling the degradation process.

Let $\{X(t),t\geq 0\}$ be an ICTMC on a state space $S=\{0,1,\ldots,M\}$, which is finite or infinite. As in the previous section, we define the transition rate as

$$\lim_{\Delta t\to 0}\frac{1-p_{ii}(t,\Delta t)}{\Delta t}=\lambda_{ii}(t), \tag{4.16}$$

$$\lim_{\Delta t\to 0}\frac{p_{ij}(t,\Delta t)}{\Delta t}=\lambda_{ij}(t), i\neq j, \tag{4.17}$$

where $\lambda_{ij}(t)$ is called the transition rate from state i to state j at time t, and $p_{ij}(t,\Delta t)$ is the transition probability of the ICTMC from state i to state j at time t. Due to the time dependency, to obtain the closed-form solutions to the ICTMC differential equations (4.18) is difficult:

$$\frac{d}{dt}p_i(t)=\sum_{\substack{j\in S\\ j\neq i}}p_j(t)\lambda_{ji}(t)-p_i(t)\sum_{\substack{j\in S\\ j\neq i}}\lambda_{ij}(t). \tag{4.18}$$

To obtain the state probability vector $\mathbf{p}(t)=(p_i(t), i\in S, t\geq 0)$, Equation (4.18) has to be solved by numerical methods. In this chapter, we introduce two kinds of numerical methods: the Runge-Kutta methods and the Monte Carlo simulation (MCS) method.

The Runge-Kutta Methods compose an important family of iterative approximation methods used to solve the differential equations of ICTMC. Let $\boldsymbol{\Lambda}(t)=(\lambda_{ij}(t), t\geq 0)$ denote the transition matrix of ICTMC. Equation (4.18) can be rewritten as

$$\frac{d}{dt}\mathbf{p}(t)=\mathbf{p}(t)\boldsymbol{\Lambda}(t)=\boldsymbol{f}(t,\mathbf{p}(t)). \tag{4.19}$$

The main idea of Runge-Kutta methods is to compute $\mathbf{p}(t+\Delta t)$ by adding to $\mathbf{p}(t)$ the product of the weighted sum of s, which derivatives at different locations within the time interval $(t, t+\Delta t)$. Mathematically, $\mathbf{p}(t+\Delta t)$ can be expressed as

$$\mathbf{p}(t+\Delta t)=\mathbf{p}(t)+\Delta t\cdot\sum_{i=1}^{s} b_i\cdot\gamma_i, \tag{4.20}$$

$$\gamma_i=\boldsymbol{f}\left(t+c_i\cdot\Delta t, \mathbf{p}(t)+\Delta t\cdot\sum_{j=1}^{s} a_{ij}\cdot\gamma_j\right), \tag{4.21}$$

where $\boldsymbol{f}$ is the first-order derivative of $\mathbf{p}(t)$ at $(t+c_i\cdot\Delta t,\ \mathbf{p}(t)+\Delta t\cdot\sum_{j=1}^{s} a_{ij}\cdot\gamma_j)$, and a_{ij}, b_i and c_i are the coefficients which are usually arranged in a Butcher Table:

$$\begin{array}{c|ccc} c_1 & a_{11} & \cdots & a_{1s} \\ c_2 & a_{21} & \cdots & a_{2s} \\ \vdots & \vdots & \ddots & \vdots \\ c_s & a_{s1} & \cdots & a_{ss} \\ \hline & b_1 & \cdots & b_s \end{array}. \tag{4.22}$$

A Runge-Kutta method is consistent if $\sum_{j=1}^{s} a_{ij}=c_i$. The Runge-Kutta method is explicit if the Butcher Table in Equation (4.22) is a lower triangular matrix; if it is not necessarily a lower triangular, then the Runge-Kutta method will be implicit, which is more general than the explicit case.

The elements in the Butcher Table are coefficients chosen to match as many of the terms in the Taylor series

$$\begin{aligned}\mathbf{p}(t+\Delta t)=&\ \mathbf{p}(t)+\Delta t\cdot\mathbf{p}^{(1)}(t)+\frac{\Delta t^2}{2!}\cdot\mathbf{p}^{(2)}(t)+\frac{\Delta t^3}{3!}\cdot\mathbf{p}^{(3)}(t)\\ &+\ldots+\frac{\Delta t^s}{s!}\cdot\mathbf{p}^{(s)}(t)+o(\Delta t^{s+1}),\end{aligned} \tag{4.23}$$

so as to minimize the approximation error. The vector quantity $\boldsymbol{p}^{(i)}$ can be expressed by $\boldsymbol{f} = \boldsymbol{p}^{(1)}$ and its derivatives, for example

$$\boldsymbol{p}^{(2)} = \frac{\partial \boldsymbol{f}}{\partial t} + \boldsymbol{f} \cdot \frac{\partial \boldsymbol{f}}{\partial \boldsymbol{p}} \text{ and } \boldsymbol{p}^{(3)} = \frac{\partial^2 \boldsymbol{f}}{\partial t^2} + 2 \cdot \boldsymbol{f} \cdot \frac{\partial^2 \boldsymbol{f}}{\partial t \partial \boldsymbol{p}} + \boldsymbol{f}^2 \cdot \frac{\partial^2 \boldsymbol{f}}{\partial \boldsymbol{p}^2} + \frac{\partial \boldsymbol{f}}{\partial \boldsymbol{p}} \cdot \left(\frac{\partial \boldsymbol{f}}{\partial t} + \boldsymbol{f} \cdot \frac{\partial \boldsymbol{f}}{\partial \boldsymbol{p}} \right).$$

On the other hand, γ_i can also be expressed by $\boldsymbol{f}$ and its derivatives, by using the Taylor series

$$\gamma_i = \boldsymbol{f}\left(t, \mathbf{p}(t)\right) + c_i \cdot \Delta t \cdot \frac{\partial \boldsymbol{f}}{\partial t}\left(t, \mathbf{p}(t)\right) + \sum_{j=1}^{s} a_{ij} \cdot \gamma_j \cdot \Delta t \cdot \frac{\partial \boldsymbol{f}}{\partial \mathbf{p}}\left(t, \mathbf{p}(t)\right) + O\left(\Delta t^3\right). \quad (4.24)$$

The coefficients in the Butcher Table (4.22) can, thus, be obtained by setting the right-hand side of Equation (4.20) equal to the Taylor series of $\mathbf{p}(t + \Delta t)$ in Equation (4.23). For example, a general form of an explicit 2-stage Runge-Kutta Method is

$$\mathbf{p}(t + \Delta t) = \mathbf{p}(t) + \Delta t \cdot \sum_{i=1}^{2} b_i \cdot \gamma_i, \quad (4.25)$$

where $\gamma_1 = \boldsymbol{f}\left(t, \mathbf{p}(t)\right)$, $\gamma_2 = \boldsymbol{f}\left(t + c_2 \cdot \Delta t, \mathbf{p}(t) + \Delta t \cdot a_{21} \cdot \gamma_1\right)$ and the coefficients are in a lower triangular Butcher Table. By Taylor expansion, we obtain

$$\gamma_2 = \boldsymbol{f}\left(t, \mathbf{p}(t)\right) + c_2 \cdot \Delta t \cdot \frac{\partial \boldsymbol{f}}{\partial t}\left(t, \mathbf{p}(t)\right) + a_{21} \cdot \Delta t \cdot \boldsymbol{k}_1 \cdot \frac{\partial \boldsymbol{f}}{\partial \mathbf{p}}\left(t, \mathbf{p}(t)\right), \quad (4.26)$$

$$\begin{aligned} \mathbf{p}(t + \Delta t) &= \mathbf{p}(t) + (b_1 + b_2) \cdot \Delta t \cdot \boldsymbol{f}\left(t, \mathbf{p}(t)\right) \\ &+ \left[c_2 \cdot b_2 \cdot \frac{\partial \boldsymbol{f}}{\partial t}\left(t, \mathbf{p}(t)\right) + a_{21} \cdot b_2 \cdot \boldsymbol{f}\left(t, \mathbf{p}(t)\right) \cdot \frac{\partial f}{\partial p}\left(t, p(t)\right) \right] \cdot \Delta t^2 + O\left(\Delta t^3\right). \end{aligned} \quad (4.27)$$

Therefore, $b_1 + b_2 = 1$, $c_2 = a_{21}$ and $c_2 \cdot b_2 = a_{21} \cdot b_2 = 1/2$.

In the MCS approach, Equation (4.18) is rewritten as

$$\frac{d}{dt} p_i(t) = \sum_{\substack{j=0 \\ j \neq i}}^{M} p_j(t) \rho_{ji}(t) \lambda_j(t) - p_i(t) \lambda_i(t), \quad (4.28)$$

where $\lambda_i(t) = \sum_{\substack{j=0 \\ j \neq i}}^{M} \lambda_{ij}(t)$ and $\rho_{ji}(t) = \lambda_{ji}(t) / \lambda_j(t)$. The quantity $\rho_{ji}(t)$ is regarded as the conditional probability that given the transition out of state j at time t, with the transition arrival state i. To rewrite Equation (4.28) in an integral form, we use an integrating factor $B_i(t) = \exp\left[\int_t^0 \lambda_i(t') dt'\right]$ is used. Multiplying both sides of Equation (4.28) by the integrating factor, one obtains

$$\frac{d}{dt}\left[p_i(t)B_i(t)\right]=B_i(t)\sum_{\substack{j=0\\j\neq i}}^{M}p_j(t)\rho_{ji}(t)\lambda_j(t). \tag{4.29}$$

Taking the integral of both sides, we have

$$\begin{aligned}
p_i(t)B_i(t)&=p_i(0)+\int_0^t\left[B_i(t')\sum_{\substack{j=0\\j\neq i}}^{M}p_j(t')\rho_{ji}(t')\lambda_j(t')\right]dt'\\
&\equiv p_i(t)=p_i(0)\exp\left[-\int_0^t\lambda_i(t')dt'\right]+\\
&\int_0^t\exp\left[-\int_{t'}^t\lambda_i(t'')dt''\right]\sum_{\substack{j=0\\j\neq i}}^{M}p_j(t')\rho_{ji}(t')\lambda_j(t')dt'.
\end{aligned} \tag{4.30}$$

In the MCS of the Markov process, the probability distribution function $p_i(t)$ is not sampled directly. Instead, the holding time at each given state i is sampled, and, then, the transition from state j to another state j is determined. This procedure is repeated until the accumulated holding time reaches the predefined time horizon. The resulting time sequence consists of the holding times at different states. To sample the holding time, the probability density (or total frequency) of departing state i, $\psi_i(t)$, can be obtained by multiplying $\lambda_i(t)$ to both sides of Equation (4.30)

$$\begin{aligned}
\psi_i(t)&=\lambda_i(t)p_i(t)\\
&=p_i(0)\lambda_i(t)\exp\left[-\int_0^t\lambda_i(t')dt'\right]\\
&+\int_0^t\lambda_i(t)\exp\left[-\int_{t'}^t\lambda_i(t'')dt''\right]\sum_{\substack{j=0\\j\neq i}}^{M}\psi_j(t')\rho_{ji}(t')dt'\\
&=p_i(0)\phi_i(t\,|\,0)+\sum_{\substack{j=0\\j\neq i}}^{M}\int_0^t\psi_j(t')\rho_{ji}(t')\phi_i(t\,|\,t')dt',
\end{aligned} \tag{4.31}$$

where

$$\phi_i(t\,|\,t')=\lambda_i(t)\exp\left[-\int_{t'}^t\lambda_i(t'')dt''\right]\quad t\geq t', \tag{4.32}$$

is defined as the conditional probability density function (pdf) that the process will depart state i at time t given that the process is at state i at time t'. Equation (4.31) indicates that the pdf $\psi_i(t)$ consists of the sum of contributions from the random walks with transitions passing through all the states (including state i) from time 0 to t. From Equation (4.31), the MCS procedure mentioned above can be derived: the cumulative distribution function (cdf) of the holding time, as shown in Equation (4.33), is obtained by integrating Equation (4.32)

$$\Phi_i(t \mid t') = 1 - \exp\left[-\int_{t'}^{t} \lambda_i(t'')dt''\right] \tag{4.33}$$

Now, given the current time t' at state i, the holding time t can be sampled through direct inversion sampling, acceptance-rejection sampling, and other sampling techniques. Following the departure from state i, the sampling of the arrived state j^* can be done by choosing a uniformly distributed random number U_0 and selecting the state which satisfies the following condition

$$\sum_{l=0}^{j^*-1} \rho_{il}(t) < U_0 < \sum_{l=0}^{j^*} \rho_{il}(t). \tag{4.34}$$

Example 4.4

We consider a non-repairable system with two pumps. Both pumps have a time-dependent failure rate $\lambda(t) = 0.3t$ where t is the system time of evolution. The failure of each pump can lead to a system crash or a safe shutdown of the other pump. The transition diagram is shown in Figure 4.3.

The corresponding transition rate matrix is

$$\Lambda(t) = \begin{bmatrix} -0.6t & 0.36t & 0 & 0.24t \\ 0 & -0.3t & 0.18t & 0.12t \\ 0 & 0 & 0 & 0 \\ 0 & 0 & 0 & 0 \end{bmatrix}$$

The numerical solution techniques are applied to this problem. Table 4.1 summarizes the probabilities of the safe and unsafe failure states at the time steps from 1 to 5.

As to the computational efficiencies, the Runge-Kutta method (time interval size = 0.1) takes 0.2899 seconds, and MCS (50,000 runs) takes 36.24 seconds to obtain the state probabilities presented in Table 4.1.

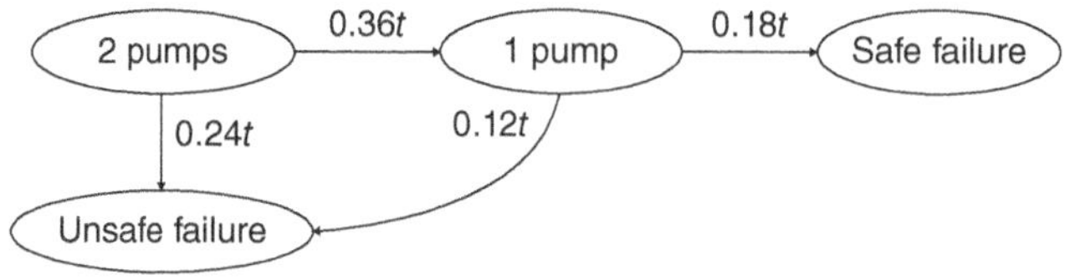

Figure 4.3 Degradation Process of a Two-pump System.

Table 4.1 Results for the degradation model obtained by the two solution techniques.

Time	Runge-Kutta method		MCS	
	Prob (safe failure)	Prob (unsafe failure)	Prob (safe failure)	Prob (unsafe failure)
1	0.0099	0.1284	0.1015	0.1221
2	0.0843	0.3497	0.0857	0.3530
3	0.2098	0.5175	0.2124	0.5142
4	0.3045	0.6004	0.3012	0.5977
5	0.3456	0.6302	0.3551	0.6331

4.3 Semi-Markov Process (SMP)

In reliability engineering, Markov chains are frequently used for modeling multi-state systems (MSSs). The sojourn time in a Markov process usually refers to the time it takes for an MSS to degrade from one state to another. In CTMCs, the sojourn time is assumed to follow an exponential distribution. However, in practical industrial systems, the degradation time typically shows a non-exponential distribution, e.g. Weibull distribution, gamma distribution, and Birnbaum-Saunders distribution, etc. Therefore, we need a more flexible model to describe such systems.

An SMP is one that changes states in accordance with a Markov chain but takes arbitrary amounts of time between changes. Let $\{X(t), t \geq 0\}$ denote an SMP defined on a state space $S = \{0, 1, \ldots, M\}$, which is finite or infinite. We decompose the SMP into several parts: denote $T = \{T_k, k \in \mathbb{N}\}$ as the successive time points when state changes in $\{X(t), t \geq 0\}$ occur $J = \{J_k, k \in \mathbb{N}\}$ as the process visited states at the corresponding time points T_k. Let $Y = \{Y_k, k \in \mathbb{N}\}$ be the successive sojourn times on the visited state, and obviously, it is $Y_k = T_{k+1} - T_k, \forall k \in \mathbb{N}$. The relationship between the process $\{X(t), t \geq 0\}$ and the process $J = \{J_k, k \in \mathbb{N}\}$ is given by

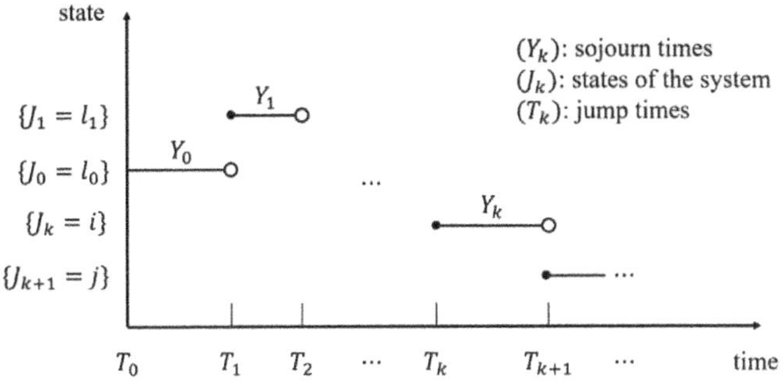

Figure 4.4 Sample Path of an SMP.

$$X(t) = X_t = J_k,\ T_k \le t < T_{k+1}, t \ge 0, k \in \mathbb{N}, \tag{4.35}$$

or equivalently,

$$X_t = J_{N(t)}, \quad \text{or} \quad J_k = X_{Z_k}, \tag{4.36}$$

where $N(t) = \max\{k \in \mathbb{N} \mid T_k \le t\}$ is the counting process of the number of jumps in $[0,t]$, $Z_k = \min\{t > T_{k-1} \mid X_k \ne X_{k-1}\}$ for all $k \in \mathbb{N}^*$ and $Z_0 = 0$ where $\mathbb{N}^*$ denotes the set of positive integers. The sample path of an SMP is demonstrated in Figure 4.4.

Thus, the following conditional independence relation holds:

$$\begin{aligned} &\Pr\left(X(t+\Delta t) = j \mid X(t) = i, X(u) = h(u), 0 \le u < t\right) \\ &= \Pr(J_{k+1} = j, T_{k+1} - T_k \le \Delta t \mid (J_0, T_0), (J_1, T_1), \ldots, (J_k = i, T_k)) \\ &= \Pr(J_{k+1} = j, T_{k+1} - T_k \le \Delta t \mid J_k = i). \end{aligned} \tag{4.37}$$

If the sojourn time at any state $(T_{k+1} - T_k, \forall k \in \mathbb{N})$ follows the exponential distribution with parameter λ, the SMP reduces to the HCTMC:

$$\begin{aligned} &\Pr\left(X(t+\Delta t) = j \mid X(t) = i, X(u) = h(u), 0 \le u < t\right) \\ &= \Pr(J_{k+1} = j, T_{k+1} - T_k \le \Delta t \mid (J_0, T_0), (J_1, T_1), \ldots, (J_k = i, T_k)) \\ &= \Pr(J_{k+1} = j, T_{k+1} - T_k \le \Delta t \mid J_k = i) \\ &= \Pr(J_{k+1} = j \mid J_k = i)\left(1 - e^{-\lambda \Delta t}\right). \end{aligned} \tag{4.38}$$

For the sojourn process, $Y = \{Y_k, k \in \mathbb{N}\}$, we define the cumulative distribution of the sojourn time as

$$H_i(t) = \Pr\left(Y_k \le t \mid J_k = i\right), \forall i \in S, k \in \mathbb{N} \tag{4.39}$$

and, the probability density distribution of the sojourn time

$$h_i(t) = \frac{d}{dt} H_i(t), \forall i \in S, k \in \mathbb{N}\mathbb{N} \tag{4.40}$$

the conditional cumulative distribution of the sojourn time

$$F_{ij}(t) = \Pr(Y_k \le t \mid J_k = i, J_{k+1} = j), \forall i, j \in S, k \in \mathbb{N} \tag{4.41}$$

and the conditional probability density distribution of the sojourn time

$$f_{ij}(t) = \frac{d}{dt} F_{ij}(t), \forall i, j \in S, k \in \mathbb{N} \tag{4.42}$$

Note that the process $J = \{J_k, k \in \mathbb{N}\}$ is the embedded Markov chain of the SMP. The transition probability matrix $\boldsymbol{P}' = (p'_{ij}, i, j \in S)$ of $J = \{J_k, k \in \mathbb{N}\}$ is constructed by

$$p'_{ij} = \Pr(J_{k+1} = j \mid J_k = i), \forall i, j \in S, k \in \mathbb{N}. \tag{4.43}$$

We define the matrix $\boldsymbol{Q}(t) = (Q_{ij}(t), i, j \in S, t \ge 0)$, with the cumulative semi-Markov kernel

$$Q_{ij}(t) = \Pr(J_{k+1} = j, Y_k \le t \mid J_k = i), \forall i, j \in S, k \in \mathbb{N} \tag{4.44}$$

and the semi-Markov kernel

$$\boldsymbol{q}(t) = \frac{d}{dt} \boldsymbol{Q}(t) \tag{4.45}$$

Now, any matrix-valued function $\boldsymbol{q}(t) = (q_{ij}(t), i, j \in S, t \ge 0)$, which satisfies the following properties, could be the semi-Markov kernel: (1) $q_{ij}(t) \ge 0, \forall i, j \in S, t \ge 0$; (2) $q_{ij}(0) = 0, \forall i, j \in S$; (3) $\int_0^\infty \sum_{j \in S} q_{ij}(t) dt = 1, \forall i \in S$.

Therefore, we obtain

$$p'_{ij} = \lim_{t \to \infty} Q_{ij}(t), \forall i, j \in S, t \ge 0 \tag{4.46}$$

and

$$f_{ij}(t) = \begin{cases} \dfrac{q_{ij}(t)}{p'_{ij}}, p'_{ij} > 0 \\ 1_{\{t=\infty\}}, p'_{ij} = 0 \end{cases}, \forall i, j \in S, t \ge 0 \tag{4.47}$$

$$h_i(t) = \sum_{j \in S} q_{ij}(t), \forall i \in S, t \ge 0 \tag{4.48}$$

Example 4.5

Consider the three-state semi-Markov system given in Figure 4.5. The distribution of the holding times between the states are $F_{12} = \exp(\lambda_1)$, $F_{21} = \text{Weibull}(\alpha_1, \beta_1)$, $F_{23} = \text{Weibull}(\alpha_2, \beta_2)$, and $F_{31} = \exp(\lambda_2)$, respectively.

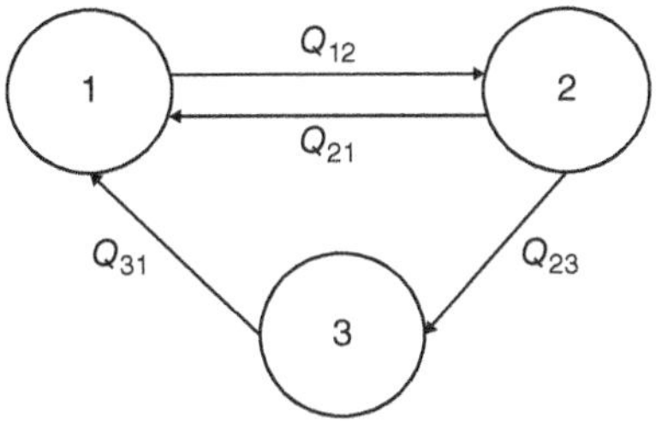

Figure 4.5 A Three-state semi-Markov System.

The transition probability matrix of the embedded Markov chain is

$$\boldsymbol{P}' = \begin{bmatrix} 0 & 1 & 0 \\ p'_{21} & 0 & p'_{23} \\ 1 & 0 & 0 \end{bmatrix},$$

where

$$p'_{21} = 1 - p'_{23} = \int_0^{\infty} \left[1 - F_{23}(t)\right] dF_{21}(t) = \int_0^{\infty} \frac{\beta_1}{\alpha_1}\left(\frac{t}{\alpha_1}\right)^{\beta_1 - 1} \exp\left(-\left(\frac{t}{\alpha_1}\right)^{\beta_1} - \left(\frac{t}{\alpha_2}\right)^{\beta_2}\right) dt.$$

Hence, we have the semi-Markov kernel of the three-state system as

$$\boldsymbol{q}(t) = \begin{bmatrix} 0 & q_{12}(t) & 0 \\ q_{21}(t) & 0 & q_{23}(t) \\ q_{31}(t) & 0 & 0 \end{bmatrix},$$

where

$$q_{12}(t) = \lambda_1 \exp(-\lambda_1 t),$$

$$q_{21}(t) = p'_{21} \frac{\beta_1}{\alpha_1}\left(\frac{t}{\alpha_1}\right)^{\beta_1 - 1} \exp\left(-\left(\frac{t}{\alpha_1}\right)^{\beta_1}\right),$$

$$q_{23}(t) = p'_{23} \frac{\beta_2}{\alpha_2}\left(\frac{t}{\alpha_2}\right)^{\beta_2 - 1} \exp\left(-\left(\frac{t}{\alpha_2}\right)^{\beta_2}\right),$$

$$q_{31}(t) = \lambda_2 \exp(-\lambda_2 t).$$

4.3.1 Markov Renewal Process

The process $(J,T)=\{(J_k,T_k),k\in\mathbb{N}\}$ is a Markov renewal chain (MRC) associated with the semi-Markov kernel $\boldsymbol{q}(t)$, if it satisfies the following condition:

$$\begin{aligned}\Pr(J_{k+1}=j,T_{k+1}-T_k\le t\mid(J_0,T_0),(J_1,T_1),\ldots,(J_k=i,T_k))\\=\Pr(J_{k+1}=j,T_{k+1}-T_k\le t\mid J_k=i)\end{aligned}\tag{4.49}$$

From Equation (4.36), we can see that the SMP is a stochastic process generated on the basis of an MRC.

We define an integral linear equation of the form

$$\varphi(i,t)=g(i,t)+\sum_{y\in S}\int_0^t q_{ij}(s)\varphi(j,t-s)ds\tag{4.50}$$

as a Markov renewal equation where $\boldsymbol{q}(t)=(q_{ij}(t),i,j\in S,t\ge 0)$ is the semi-Markov kernel, $g(i,t)$ is a given function defined on $S\times\mathbb{R}_{\ge 0}$, where $\mathbb{R}_{\ge 0}$ denotes the set of positive real numbers, and φ is the unknown function. Let us denote $\boldsymbol{\varphi}(t)=(\varphi(i,t),i\in S)$ and $\mathbf{g}(t)=(g(i,t),i\in S)$; then

$$\boldsymbol{\varphi}(t)=\mathbf{g}(t)+\boldsymbol{q}(t)*\boldsymbol{\varphi}(t),\tag{4.51}$$

where * denotes the convolution product. And Equation (4.51) is equivalent to

$$(\delta\boldsymbol{I}-\boldsymbol{q}(t))*\boldsymbol{\varphi}(t)=g(t),\tag{4.52}$$

where δ is the Kronecker delta.

We define the transition probability matrix of the SMP as the matrix-valued function $\boldsymbol{P}(t)=(p_{ij}(t),i,j\in S,t\ge 0)$, with

$$\begin{aligned}p_{ij}(t)&=\Pr(X(s+t)=j\mid X(s)=i)\\&=\Pr(X(t)=j\mid X(0)=i),\forall i,j\in S,t\ge 0.\end{aligned}\tag{4.53}$$

and we write

$$\begin{aligned}p_{ij}(t)&=\Pr(X(t)=j|X(0)=i)\\&=\Pr(X(t)=j,T_1>t|X(0)=J_0=i)\\&+\Pr(X(t)=j,T_1\le t|X(0)=J_0=i)\\&=\delta_{ij}\Pr(Y_1>t)+E_{J_1,T_1}\{(T_1<t)\Pr(X(t)=j\mid(J_1,T_1))\}\\&=\delta_{ij}(1-H_i(t))+\sum_{l\in S}\int_0^t dQ_{il}(s)p_{lj}(t-s),\end{aligned}\tag{4.54}$$

where δ_{ij} is the Kronecker delta, and $E\{\cdot\}$ denotes the expectation. Equation (4.54) can be then, equivalently written in matrix form:

$$\boldsymbol{P}(t) = \boldsymbol{I} - \boldsymbol{H}(t) + \boldsymbol{q}(t) * \boldsymbol{P}(t). \tag{4.55}$$

Similarly, we can obtain

$$\boldsymbol{F}(t) = \boldsymbol{Q}(t) + \boldsymbol{q}(t) * \boldsymbol{F}(t), \tag{4.56}$$

where $\boldsymbol{F}(t) = \left(F_{ij}(t), i, j \in S, t \geq 0\right)$ denotes the matrix of the conditional cumulative distribution of sojourn time of the SMP. The problem of deriving the transition probability matrix or the conditional cumulative distribution of sojourn time of the SMP is equivalent to solve the corresponding Markov renewal Equations (4.55) and (4.56).

Example 4.6

We consider a system with state space $S = \{i \in \mathbb{N} \mid 0 \leq i \leq s\}$, and consider a partition of S into two non-empty sets, S_0 and S_1, where S_0 contains all the functioning states and S_1 contains all the failure states. Set $|S_0| = r$ and $|S_1| = s - r$ where $|\cdot|$ denotes the cardinal number of the set. Let $\boldsymbol{q}(t) = (q_{ij}(t), i, j \in S, t \geq 0)$ denote the semi-Markov kernel of the system and the initial state probability vector $\boldsymbol{\alpha} = \mathbf{p}(0) = \left(p_i(0), i \in S\right)$. Now, consider the following partition of the semi-Markov kernel:

$$\boldsymbol{q}(t) = \begin{bmatrix} \boldsymbol{q}_{00}(t) & \boldsymbol{q}_{01}(t) \\ \boldsymbol{q}_{10}(t) & \boldsymbol{q}_{11}(t) \end{bmatrix},$$

and the partition of the initial state probability vector:

$$\boldsymbol{\alpha} = \begin{bmatrix} \alpha_0 & \alpha_1 \end{bmatrix}.$$

The reliability of the system at time t, $R(t)$, can, then, be expressed by the probability of the event $\{\forall u \in [0,t], X(u) \in S_0\}$, i.e.

$$\begin{aligned} R(t) &= \Pr\left(\forall u \in [0,t], X(u) \in S_0\right) \\ &= \Pr\left(W(t) \in S_0\right) \\ &= \sum_{j \in S_0} \sum_{i \in S_0} \Pr(W(t) = j \mid W(0) = i) \Pr\left(W(0) = i\right) \\ &= \sum_{j \in S_0} \sum_{i \in S_0} \alpha(i) p_{ij}(t), \end{aligned}$$

where $\{W(t), t \geq 0\}$ is an SMP with state space $S_0 \cup \{\Delta\}$, Δ is an absorbing state. To state it more clearly, let T_{S_1} be the hitting time of S_1 by the process $\{X(t), t \geq 0\}$, then $W(t)$ shall satisfy

$$W(t) = \begin{cases} X(t), & t < T_{S_1} \\ \Delta, & t \geq T_{S_1} \end{cases},$$

and the semi-Markov kernel of $\{W(t), t \geq 0\}$ is

$$\begin{bmatrix} \boldsymbol{q}_{00}(t) & \boldsymbol{q}^0(t) \\ \boldsymbol{0} & 0 \end{bmatrix},$$

where $\boldsymbol{q}^0(t) = \boldsymbol{q}_{01}(t)\boldsymbol{I}_{s-r}$ and where $\boldsymbol{I}_{s-r}$ denotes an $(s-r)$-order identity matrix. Thus, we have

$$R(t) = \alpha_0 \boldsymbol{P}_{00}(t)1_r = \alpha_0\left(\boldsymbol{I} - \boldsymbol{H}_0(t)\right)1_r + \alpha_0\left(\boldsymbol{q}_{00}(t) * \boldsymbol{P}_{00}(t)\right)1_r,$$

where $\mathbf{1}_r$ denotes a r-dimensional unit column vector, and $\boldsymbol{P}_{00}(t)$ denotes the corresponding partition of the transition probability matrix. Similarly, the maintainability of the system , which indicates the probability that the system will be restored to the state of functioning within a time period *t*, if it fails, satisfies

$$M(t) = 1 - \alpha_1 \boldsymbol{P}_{11}(t)\mathbf{1}_{s-r} = \alpha_1\left(\boldsymbol{I} - \boldsymbol{H}_1(t)\right)\mathbf{1}_{s-r} + \alpha_1\left(\boldsymbol{q}_{11}(t) * \boldsymbol{P}_{11}(t)\right)1_{s-r}.$$

We consider the partition of the mean sojourn time as $\boldsymbol{m} = (\boldsymbol{m}_0, \boldsymbol{m}_1)^T$ where $(\cdot)^T$ denotes the transpose matrix. The stationary distribution of the SMP is

$$\pi = \frac{1}{\upsilon \boldsymbol{m}} diag(\boldsymbol{v})\boldsymbol{m},$$

and the steady-state availability of the system is

$$A = \frac{1}{\upsilon \boldsymbol{m}} \boldsymbol{m}^T diag(\boldsymbol{v})\mathbf{1}_s,$$

where υ denotes the stationary distribution of the embedded Markov chain $J = \{J_k, k \in \mathbb{N}\}$ and $diag(\cdot)$ denotes the diagonal matrix. The mean time to failure (MTTF) and mean time to repair (MTTR) of the system are, respectively,

$$MTTF = \alpha_0\left(\boldsymbol{I} - \boldsymbol{P}_{00}\right)^{-1} \boldsymbol{m}_0,$$

$$MTTR = \alpha_1\left(\boldsymbol{I} - \boldsymbol{P}_{11}\right)^{-1} \boldsymbol{m}_1.$$

To obtain the aforementioned functions, in this chapter, we have introduced the Laplace transform of the Markov renewal equations. Take Equation (4.51) as an example, let us define $\mathbf{G}(t) = (g_i(t), i \in S, t \geq 0)$, where $g_i(t) = 1 - H_i(t)$. Because

$$(g_i(s)) = \int_0^\infty e^{-st}\left(1 - \sum_{l \in s} Q_{il}(t)\right)dt = \frac{1}{s}\left(1 - \sum_{l \in s} Q_{il}^e(s)\right), \tag{4.57}$$

we can take the Laplace transform of Equation (4.51) and derive

$$\frac{1}{s}\boldsymbol{P}^e(s)=\boldsymbol{G}^e(s)+\frac{1}{s}\boldsymbol{Q}^e(s)\boldsymbol{P}^e(s). \tag{4.58}$$

Hence,

$$\frac{1}{s}\boldsymbol{P}^e(s)=\left[\boldsymbol{I}-\boldsymbol{Q}^e(s)\right]^{-1}\boldsymbol{G}^e(s)=\boldsymbol{U}^e(s)\boldsymbol{G}^e(s), \tag{4.59a}$$

$$\boldsymbol{P}(t)=\int_0^t d\boldsymbol{U}(u)\boldsymbol{G}(t-u). \tag{4.59b}$$

where $\boldsymbol{U}(t)=\left(U_{ij}(t),i,j\in S,t\geq 0\right)=\sum_{n=0}^{\infty}\boldsymbol{Q}^{(n)}(t)$ and $[\cdot]^{(n)}$ denotes the nth power of the matrix. Then, $U_{ij}(t)$ denotes the expected number of visits to state j in the time interval $[0,t]$, given that the process starts from state i. We call $U(t)$ the Markov renewal functions.

Example 4.7

We continue with the three-state semi-Markov system given in Figure 4.5. Suppose that states 1 and 2 are the working states, and state 3 represents the failure state, which means we have the partition $S_0=\{1,2\}$ and $S_1=\{3\}$. By solving the Markov renewal equations, we obtain

$$R(t)=\boldsymbol{\alpha}_0\left(\boldsymbol{I}-\boldsymbol{Q}_{00}(t)\right)^{-1}*\left(\boldsymbol{I}-\boldsymbol{H}_0(t)\right)1_r,$$

$$M(t)=1-\boldsymbol{\alpha}_1\left(\boldsymbol{I}-\boldsymbol{Q}_{11}(t)\right)^{-1}*\left(\boldsymbol{I}-\boldsymbol{H}_1(t)\right)1_{s-r}.$$

Therefore, we obtain

$$R(t)=\alpha(1)\left(1-Q_{21}*Q_{12}\right)^{-1}\left(1-Q_{12}*Q_{23}\right)(t)+\alpha(2)\left(1-Q_{21}*Q_{12}\right)^{-1}\left(1-Q_{23}\right)(t),$$

$$M(t)=1-\alpha(3)\left(1-Q_{31}(t)\right).$$

Assume we have the stationary distribution of the embedded Markov chain as $\upsilon=\frac{1}{2+p}(1,1,p)$ and the mean sojourn time of the SMP as $\boldsymbol{m}=(m_1,m_2,m_3)^T$; then, we obtain

$$m_1=\int_0^{\infty}\left(1-Q_{21}(t)\right)dt,$$

$$m_2=\int_0^{\infty}\left(1-Q_{21}(t)-Q_{23}(t)\right)dt,$$

$$m_3 = \int_0^{\infty} \left(1 - Q_{31}(t)\right) dt,$$

and the stationary distribution

$$\pi = (\pi_1, \pi_2, \pi_3) = \frac{1}{m_1 + m_2 + pm_3}(m_1, m_2, pm_3).$$

Thus, the steady-state availability is obtained as

$$A = \frac{m_1 + m_2}{m_1 + m_2 + pm_3}.$$

The MTTF and MTTR are, respectively,

$$MTTF = \frac{1}{1 - p'_{21}}(m_1 + m_2)$$
$$MTTR = m_3.$$

4.4 Piecewise Deterministic Markov Process (PDMP)

PDMP is adopted to treat the system dynamics and the degradation dependence in multi-state physics systems. For this, the degradation processes are classified into two groups: (1) $\boldsymbol{L} = \{L_1, L_2, \ldots, L_M\}$ modeled by M physics-based models and (2) $\boldsymbol{K} = \{K_1, K_2, \ldots, K_N\}$ modeled by N multi-state models (MSMs) where $L_m, m = 1, 2, \ldots, M$ and $K_n, n = 1, 2, \ldots, N$ are the indexes of the degradation processes. Let $\overrightarrow{X_{L_m}}(t)$ denote the time-dependent continuous variables of the degradation process L_m and $Y_{K_n}(t)$ denote the state variable of the degradation process K_n.

Dependence between degradation processes may exist within each group and between the two groups, for example, the evolution of $\overrightarrow{X_{L_m}}(t)$ may be influenced by the degradation states of $\overrightarrow{X_{m'}}(t), m \neq m'$ and $Y_{K_n}(t)$;the transition rates of $Y_{K_n}(t)$ may be influenced by the degradation states of $Y_{K_{n'}}(t), n \neq n'$ and $\overrightarrow{X_{L_m}}(t)$. An illustration of a system with two dependent degradation processes is shown in Figure 4.6 where the further degraded states of $K_1(L_1)$ lead to higher degradation rates of L_1 (higher transition rates of K_1 to step to further degraded states). In this particular case, the degradation rate of $\overrightarrow{X_{L_1}}(t)$ changes at the same time when $Y_{K_1}(t)$ changes. However, this does not necessarily occur in all cases because the degradation rate of $\overrightarrow{X_{L_1}}(t)$ may also depend on other influencing factors and the related coefficients in the physics equations.

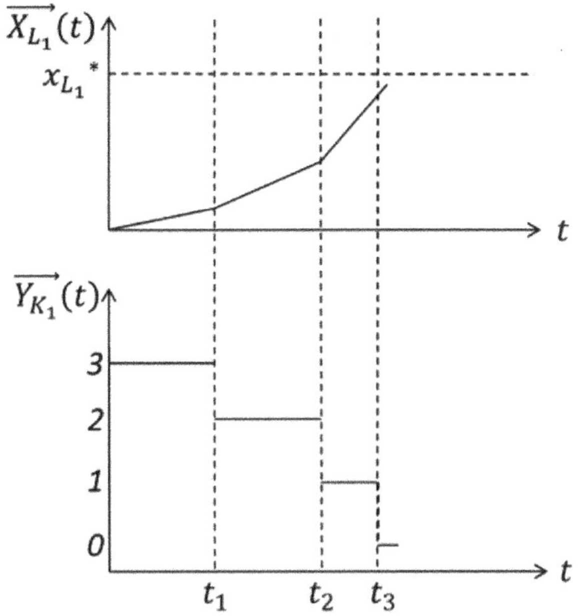

Figure 4.6 An illustrative example of a system with two dependent degradation processes. (Top Figure: degradation process of L_1; Bottom Figure: degradation process of K_1).

Let

$$\vec{Z}(t)=\begin{pmatrix}\begin{pmatrix}\overrightarrow{X_{L_1}}(t)\\ \vdots\\ \overrightarrow{X_{L_M}}(t)\end{pmatrix}=\vec{X}(t)\\ \begin{pmatrix}Y_{K_1}(t)\\ \vdots\\ Y_{K_N}(t)\end{pmatrix}=\vec{Y}(t)\end{pmatrix}\in \boldsymbol{E}=\mathbb{R}^{d_L}\times \boldsymbol{S} \tag{4.60}$$

denote the overall degradation processes of the system where $\boldsymbol{E}$ is the space combining $\mathbb{R}^{d_L}$ ($d_L=\sum_{m=1}^{M} d_{L_m}$) and $\boldsymbol{S}$. The evolution of $\vec{Z}(t)$ involves two parts:

1) The stochastic behavior of $\vec{Y}(t)$, which is governed by the transition rates depending on the degradation states of all the degradation processes in the system:

$$\lim_{\Delta t \to 0} P\left(\vec{Y}(t+\Delta t)=\vec{j}\,\middle|\,\overrightarrow{X}(t),\vec{Y}(t)=\vec{i},\theta_K=\bigcup_{n=1}^{N}\theta_{K_n}\right)/\Delta t$$
$$=\lambda_{\vec{i}}\left(\vec{j}\mid\overrightarrow{X}(t),\theta_K\right),\forall t\geq 0,\vec{i},\vec{j}\in\Sigma,\vec{i}\neq\vec{j}. \tag{4.61}$$

2) The deterministic behavior of $\overrightarrow{X}(t)$ between two consecutive jumps of $\vec{Y}(t)$, which is described by the deterministic physics equations depending on the degradation states of all the degradation processes in the system

$$\dot{\vec{X}}(t)=\begin{pmatrix}\overrightarrow{\dot{X}_{L_1}}(t)\\ \vdots \\ \overrightarrow{\dot{X}_{L_M}}(t)\end{pmatrix}=\begin{pmatrix}\overrightarrow{f_{L_1}^{\vec{Y}(t)}}(\overrightarrow{X}(t),t\mid\theta_{L_1})\\ \vdots \\ \overrightarrow{f_{L_M}^{\vec{Y}(t)}}(\overrightarrow{X}(t),t\mid\theta_{L_M})\end{pmatrix}$$
$$=\overrightarrow{f_L^{\vec{Y}(t)}}(\overrightarrow{X}(t),t\mid\theta_L=\bigcup_{m=1}^{M}\theta_{L_m}). \tag{4.62}$$

Let T_k denote the k-th transition time of the process $\vec{Y}(t)$. The set $\left\{\overrightarrow{Z_k},\ T_k\right\}_{k\geq 0}$ is, then, a Markov renewal process [10] defined on the space $\boldsymbol{E}\times\mathbb{R}^{+}$. The probability that the whole system will step to state $\vec{j}$ from state $\vec{i}$, $\vec{i},\vec{j}\in\boldsymbol{E}$, $\vec{i}\neq\vec{j}$ in the time interval$\left[T_n,\ T_n+\Delta t\right]$, given $\left\{\overrightarrow{Z_k},\ T_k\right\}_{k\leq n}$, is:

$$P\left[\overrightarrow{Z_{n+1}}=\vec{j},\ T_{n+1}\in\left[T_n,\ T_n+\Delta t\right]\mid\left\{\overrightarrow{Z_k},\ T_k\right\}_{k\leq n-1},\left\{\overrightarrow{Z_n}=\vec{i},\ T_n\right\}\right]$$
$$=P\left[\overrightarrow{Z_{n+1}}=\vec{j},\ T_{n+1}\in\left[T_n,\ T_n+\Delta t\right]\mid\overrightarrow{Z_n}=\vec{i}\right],\forall n\geq 0,\ \vec{i},\vec{j}\in\boldsymbol{E},\ \vec{i}\neq\vec{j}. \tag{4.63}$$

Let $\mathcal{F}$ denote the predefined space of the failure states of $\vec{Z}(t)$; then, the system reliability at time t is defined as

$$R(t)=P\left[\vec{Z}(s)\notin\mathcal{F},\forall s\leq t\right]. \tag{4.64}$$

To consider a general setting, $\mathcal{F}$ is dependent on system topology, which is problem-specific and can be determined by using reliability analysis tools, such as fault tree analysis.

For reliability assessment, MCS and the finite-volume (FV) method are two widely used numerical approaches to solve PDMP. First, we illustrate a detailed description of the procedures of the MCS method. We rewrite Equation (4.63) as

$$P\left[\overrightarrow{Z_{n+1}} \in B,\ T_{n+1} \in \left[T_n,\ T_n + \Delta t\right] \mid \overrightarrow{Z_n} = \vec{i},\ \theta_{\boldsymbol{K}}\right]$$
$$= \iint\limits_{B^*[0,\Delta t]} N\left(\vec{i}, \overrightarrow{dz}, ds \mid \theta_{\boldsymbol{K}}\right), \forall n \geq 0, \Delta t \geq 0,\ \vec{i} \in \boldsymbol{E}, B \in \varepsilon, \tag{4.65}$$

where B is a measurable set on $\boldsymbol{E}$, ε is a σ-algebra of $\boldsymbol{E}$ [10], and $N\left(\vec{i}, \overrightarrow{dz}, ds \mid \theta_{\boldsymbol{K}}\right)$ is a semi-Markov kernel on $\boldsymbol{E}$, which verifies that $\iint\limits_{E^*[0,\Delta t]} N(\vec{i}, \overrightarrow{dz}, ds \mid \theta_{\boldsymbol{K}}) \leq 1, \forall \Delta t \geq 0,\ \vec{i} \in \boldsymbol{E}.$ It can be further developed as

$$N\left(\vec{i}, \overrightarrow{dz}, ds \mid \theta_K\right) = dF_{\vec{i}}\left(s \mid \theta_K\right) \beta\left(\vec{i}, s, \overrightarrow{dz} \mid \theta_{\boldsymbol{K}}\right)$$
(4.66a)

where

$$NdF_{\vec{i}}\left(s \mid \theta_{\boldsymbol{K}}\right)$$
(4.66b)

is the pdf of $T_{n+1} - T_n$ given $\overrightarrow{Z_n} = \vec{i}$ and

$$\beta\left(\vec{i}, s, \overrightarrow{dz} \mid ,\ _{\boldsymbol{K}}\right)$$
(4.66c)

is the conditional probability of state $\overrightarrow{Z_{n+1}}$ given $T_{n+1} - T_n = s$.

Then, the MCS method can be used to estimate the reliability of the system within a certain mission time T_{miss}, given the initial system state $\overrightarrow{Z_0}$ at time $T_0 = 0$. The method to simulate the behavior of the system consists of sampling the transition time from Equation (4.66b) and the arrival state from Equation (4.66c) for the components in the second group and, then, using the physics Equation (4.62) to calculate the evolution of the components in the first group within the transition times. Each simulation trial continues until the time of system evolution reaches T_{miss} or until the system enters the failure space $\mathcal{F}$, Afterwards, the occurrence of the simulation trial is recorded for the statistical estimation of the system reliability.

The procedure of the MCS method [11] is as follows:

Set N_{max} (the maximum number of replications) and $k = 0$ (index of MCS trials)
Set $k' = 0$ (number of MCS trials that end in failure state)
While $k < N_{max}$

Initialize the system by setting $\vec{Z}' = \begin{pmatrix} \vec{X}(0) \\ \vec{Y} \end{pmatrix}$ (initial system state) and the time

$T = 0$ (initial system time)

Set $t' = 0$ (state holding time)
While $T < T_{miss}$
Sample a random value of t' from the pdf Equation (4.66b)
Sample an arrival state $\vec{Y}'$ for stochastic process $\vec{Y}(t)$ from all possible states, by using the conditional probability function Equation (4.66c)
Set $T = T + t'$
Calculate $\vec{X}(t)$ in the interval $[T - t', T]$ by using the physics equations Equation (4.62)

Set $\vec{Z}' = \begin{pmatrix} \vec{X}(T) \\ \vec{Y}' \end{pmatrix}$

If $T \le T_{miss}$

If $\exists t \in [T - t', T]$, $\vec{Z}(t) = \begin{pmatrix} \vec{X}(t) \\ \vec{Y} \end{pmatrix} \in \mathcal{F}$

Set $k' = k' + 1$
Break
End if
Else (when $T > T_{miss}$)

If $\exists t \in [T - t',\ T_{miss}], \vec{Z}(t) = \begin{pmatrix} \vec{X}(t) \\ \vec{Y} \end{pmatrix} \in \mathcal{F}$

Set $k' = k' + 1$
Break
End if
End if
Set $\vec{Y} = \vec{Y}'$
End While
Set $k = k + 1$
End While

The estimated component reliability at time T_{miss} can be obtained as

$$\hat{R}\left(T_{miss}\right)=1-\frac{k'}{N_{max}} \tag{4.67}$$

where k' represents the number of trials that end in the failure state of the system, and the sample variance is [12]

$$var_{\hat{R}\left(T_{miss}\right)}=\frac{\hat{R}\left(T_{miss}\right)\left(1-\hat{R}\left(T_{miss}\right)\right)}{N_{max}-1}. \tag{4.68}$$

A FV scheme discretizing the state space of the continuous variables and the time space of PDMP is an alternative that can, in certain cases, lead to results comparable to the MCS method but in significantly shorter computing times. Here, we illustrate an explicit FV scheme for system reliability estimation [13].

The FV method can be applied under the following assumptions:

- The transition rates $\lambda_{\vec{i}}\left(\vec{j}\mid\cdot,\ \theta_{K}\right), \forall\vec{i},\vec{j}\in\boldsymbol{S}$ are continuous and bounded functions from $\mathbb{R}^{d_L}$ to $\mathbb{R}^{+}$.
- The physic equations $\overrightarrow{f_L^{\vec{i}}}\left(\cdot,\cdot\mid\theta_L\right), \forall\vec{i}\in\boldsymbol{S}$ are continuous functions from $\mathbb{R}^{d_L}\times\mathbb{R}^{+}$ to $\mathbb{R}^{d_L}$ and locally Lipschitz continuous.
- The physics equations $\overrightarrow{f_L^{\vec{i}}}\left(\cdot,\ t\mid\theta_L\right), \forall\vec{i}\in\boldsymbol{S}$ are sub-linear, i.e. there are some $V_1>0$ and $V_2>0$ such that

$$\forall\vec{x}\in\mathbb{R}^{d_L}, t\in\mathbb{R}^{+}\left|\overrightarrow{f_L^{\vec{i}}}(\vec{x},t\mid\theta_L)\right|\leq V_1\left(\vec{x}+|t|\right)+V_2.$$

- The functions $div\left(\overrightarrow{f_L^{\vec{i}}}\left(\cdot,\cdot\mid\theta_L\right)\right), \forall\vec{i}\in\boldsymbol{S}$ are almost everywhere bounded in absolute value by some real value $D>0$ (independent of i).

Let $\overrightarrow{\mathbf{g}^{\vec{i}}}\left(\cdot,\cdot\right):\ \mathbb{R}^{d_L}\times\mathbb{R}\to\mathbb{R}^{d_L}$ denote the solution of

$$\frac{\partial}{\partial t}\overrightarrow{\mathbf{g}^{\vec{i}}}\left(\vec{x},t\mid\theta_L\right)=\overrightarrow{f_L^{\vec{i}}}(\overrightarrow{\mathbf{g}^{\vec{i}}}\left(\vec{x},t\mid\theta_L\right),t\mid\theta_L),\forall\vec{i}\in\boldsymbol{S},\ \vec{x}\in\mathbb{R}^{d_L},t\in\mathbb{R}$$

with

$$\overrightarrow{\mathbf{g}^{\vec{i}}}\left(\vec{x},0\mid\theta_L\right)=\vec{x},\forall\vec{i}\in\boldsymbol{S},\ \to\in\mathbb{R}^{d_L},$$

where $\overrightarrow{\mathrm{g}^{\vec{i}}}\left(\vec{x},t\mid\theta_L\right)$ represents the deterministic evolution of $\vec{X}\left(t\right)$ at time t, starting from the condition $\vec{x}$ and while the processes $\vec{X}\left(t\right)$ are holding in state $\vec{i}$.
The state space $\mathbb{R}^{d_L}$ of continuous variables $X\left(t\right)$ is divided into an admissible mesh $\mathcal{M}$, which is a family of measurable subsets of $\mathbb{R}^{d_L}$ ($\mathcal{M}$ is a partition of $\mathbb{R}^{d_L}$), such that

- $\bigcup_{A \in \mathcal{M}} A = \mathbb{R}^{d_L}$;
- $\forall A, B \in \mathcal{M}$, $A \neq B \Rightarrow A \bigcap B = \varnothing$;
- $m_A = \int_A \overrightarrow{dx} > 0, \forall A \in \mathcal{M}$ where m_A is the volume of grid A; and
- $sup_{A \in \mathcal{M}} diam(A) < +\infty$ where $diam(A) = sup_{\forall \vec{x}, \vec{y} \in A} |\vec{x} - \vec{y}|$.

Additionally, the time space $\mathbb{R}^+$ is divided into small intervals $\mathbb{R}^+ = \bigcup_{n=0,1,2,\ldots} [n\Delta t, (n+1)\Delta t[$ by setting the time step $\Delta t > 0$ (the length of each interval).

Let $p_t(d\vec{z} \mid \theta = \theta_L \bigcup \theta_K)$ denote the probability distribution of $\vec{Z}(t)$. The numerical scheme aims at constructing an approximate value $\rho_t(\vec{x}, \cdot \mid \theta)\overrightarrow{dx}$ for $p_t(\overrightarrow{dx}, \cdot \mid \theta)$ so $\rho_t(\vec{x}, \cdot \mid \theta)$ is constant in each $[n\Delta t, (n+1)\Delta t[\times A \times \{y_i\}, \forall A \in \mathcal{M}, \vec{i} \in \mathbf{S}$:

$$\rho_t(\vec{x}, \vec{i} \mid \theta) = P_n(A, \vec{i} \mid \theta), \forall \vec{i} \in \mathbf{S}, \vec{x} \in A, t \in [n\Delta t, (n+1)\Delta t[. \tag{4.69}$$

$P_0(A, \vec{i} \mid \theta), \forall \vec{i} \in \mathbf{S}, A \in \mathcal{M}$ is defined as follows:

$$P_0(A, \vec{i} \mid \theta) = \frac{\int_A p_0(\overrightarrow{dx}, \vec{i} \mid \theta)}{m_A}. \tag{4.70}$$

Then, $P_{n+1}(A, \vec{i} \mid \theta), \forall \vec{i} \in \mathbf{S}, A \in \mathcal{M}, n \in \mathbb{N}$ can be calculated considering the deterministic evaluation of $\vec{X}(t)$ and the stochastic evolution of $\vec{Y}(t)$ based on $P_n(\mathcal{M}, \vec{i} \mid ,)$ by the Chapman-Kolmogorov forward equation as follows:

$$\begin{aligned} & P_{n+1}(A, \vec{i} \mid \theta) \\ & = \frac{1}{1 + \Delta t b_A^{\vec{i}}} \widehat{P_{n+1}}(A, \vec{i} \mid \theta) + \Delta t \sum_{\vec{j} \in S} \frac{a_A^{\vec{j}\vec{i}}}{1 + \Delta t b_A^{\vec{j}}} \widehat{P_{n+1}}(A, \vec{j} \mid \theta), \end{aligned} \tag{4.71}$$

where

$$a_A^{\vec{j}\vec{i}} = \int_A \lambda_{\vec{j}}(\vec{i}, \vec{x} \mid \theta_K) \overrightarrow{dx} / m_A, \forall \vec{i} \in \mathbf{S}, A \in \mathcal{M}$$

is the average transition rate from state $\vec{j}$ to state $\vec{i}$ for grid A,

$$b_A^{\vec{i}} = \sum_{\vec{j} \neq \vec{i}} a_A^{\vec{i}\vec{j}}, \forall \vec{i} \in \mathbf{S}, A \in \mathcal{M}$$

is the average transition rate out of state $\vec{i}$ for grid A,

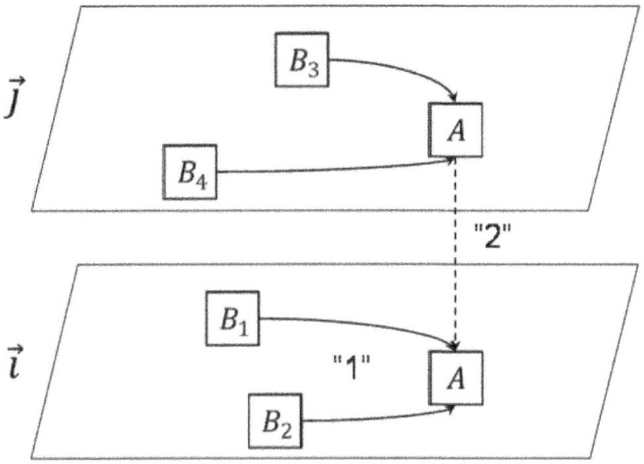

Figure 4.7 The Evolution of Degradation Processes during $\left[n\Delta t,(n+1)\Delta t\right]$ [14].

$$\widehat{P_{n+1}}\left(A,\, \vec{i} \mid \theta\right) = \sum_{B \in \mathcal{M}} m_{BA}^{\vec{i}} P_n\left(B,\, \vec{i} \mid \theta\right) / m_A, \forall \vec{i} \in \mathbf{S}, A \in \mathcal{M}$$

is the approximate value of the pdf on $\{\vec{i}\} \times \left[(n+1)\Delta t,\, (n+2)\Delta t\right] \times A$ according to the deterministic evolution of $\overrightarrow{X}(t)$, and

$$m_{BA}^{\vec{i}} = \int_{\{\vec{y} \in B \mid \overrightarrow{g^{\vec{i}}}(\vec{y}, \Delta t \mid \theta_L) \in A\}} \overrightarrow{dy}, \forall \vec{i} \in \mathbf{S}, A, B \in \mathcal{M}$$

is the volume of the part of grid B, which will enter grid A after time Δt according to the deterministic evolution of $\overrightarrow{X}(t)$.

Figure 4.7 shows an illustrative example in $\mathbb{R}^2$ to explain the procedure of FV modeling scheme.

The FV scheme solves the PDMP by considering two different situations to calculate the probabilities that $\vec{Z}(t) \in \left(\mathrm{A},\, \vec{i}\right)$, where $\forall \vec{i} \in S, A \in \mathcal{M}$ at time $(n+1)\Delta t$, according to Equation (4.71). The first one (denoted by "1" in Figure 4.7) is that $\vec{X}(t)$ evolves but $\vec{Y}(t)$ does not change, which is quantified by the first term of the right-hand part of Equation (4.71), where $\dfrac{1}{1+\Delta t b_A^{\vec{i}}}$ is the approximated probability that no transition occurs from state $\vec{i}$ for grid A, and, B_1, B_2 are the grids of which some parts will enter grid A, due to the deterministic evolution of $\overrightarrow{X}(t)$ at time $(n+1)\Delta t$, given $\vec{Y}(t) = \vec{i}$. The second one (denoted by "2" in Figure 4.7), is that $\vec{Y}(t)$ steps to state $\vec{i}$ from another state $\vec{j} \in S$, which is quantified by the second term of the right-hand part of Equation (4.71), where $a_A^{\vec{j}\vec{i}} \Delta t$ is the transition probability from state $\vec{j}$ to state $\vec{i}$ for grid A, and, B_3, B_4 are the grids of which some parts will enter grid A, due to the deterministic evolution of $\overrightarrow{X}(t)$ at time $(n+1)\Delta t$, given $\vec{Y}(t) = \vec{j}$.

The approximate solution $\rho_t\left(\vec{x},\cdot|\theta\right)d\vec{x}$ weakly converges toward $p_t\left(d\vec{x},\cdot|\theta\right)$ when $\Delta t \to 0$ and $\left|\mathcal{M}\right| / \Delta t \to 0$ where $\left|\mathcal{M}\right| = sup_{A\in\mathcal{M}}\, diam\left(A\right)$.

The reliability of the system can, then, be calculated as

$$R\left(t\right) = \int_{\vec{z}\notin\mathcal{F}} p_t\left(d\vec{z}\,|\,,\,\right). \tag{4.72}$$

4.5 Exercises

1 Consider a job shop consisting of M machines and a single repairman. Suppose the amount of time a machine runs before breaking down is exponentially distributed with rate λ, and the amount of time it takes the repairman to fix any broken machine is exponentially distributed with rate μ. If we say the state is i whenever there are i machines down, then

a. Calculate the steady state probability distribution of the job shop.
b. Calculate the average number of machines not in use.
c. Compute the long-run proportion of time that a given machine is working.

2 Consider a system with a total of $n+1$ pumps, one of which is in use and n of which are spare pumps. When the pump in use fails and spare pumps are available, the failed pump is replaced by a spare pump immediately. The spare pumps will not fail when not in use. The failed pump will be repaired successfully. Only one pump is repaired at a time. Therefore, the state space of the system is expressed as $S = \left\{n+1, n, \ldots, 1, 0\right\}$, representing the total number of pumps not failed. The transition rate in this Markov chain model is composed of the pump failure rate function and repair rate function. The failure rate function is $\lambda\left(t\right) = \alpha_0\beta_0 t^{\beta_0 - 1}$, and the repair rate function is $\mu\left(t\right) = \alpha_1\beta_1 t^{\beta_1 - 1}$. Derive the reliability function of the system. If the mean number of failures during the time interval $\left[0, t\right]$ with initial state i is $N_i\left(t\right) = V_i\left(t\right)$, derive the reward matrix r.

3 Consider a series-parallel system as shown in Figure 4.8. In this system, the three components are mutually stochastic independent. Components 1 and 2 have two different states, respectively, and Component 3 has three different states.

The performance of the ith component at state j is denoted as g_{ij}. The probability that the ith component is at state j at time t is denoted as $p_{ij}\left(t\right)$. The transition rate of the ith component from state j to state k at time t is denoted as $\lambda_{jk}^{i}\left(t\right)$. The values of g_{ij}, $p_{ij}\left(0\right)$, and $\lambda_{jk}^{i}\left(t\right)$ are listed in the table below.

If the system demand is 1.8 ton/min, calculate the system reliability function and the mean lifetime of the system.

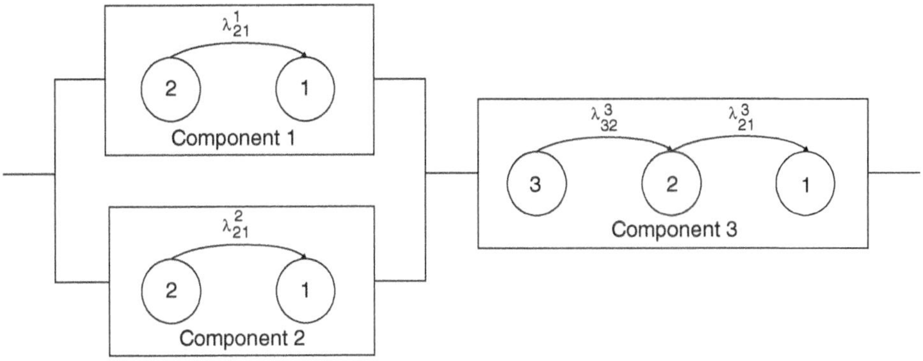

Figure 4.8 The Diagram of the Series-parallel System.

Component	State	Performance g_{ij} (ton/min)	Initial state probability $p_{ij}(0)$	Transition rate $\lambda^i_{jk}(t)$ (year) (-1)
1	1	0	0	$\lambda^1_{21}(t) = 0.8 + 0.2t$
	2	1.5	1	
2	1	0	0	$\lambda^2_{21}(t) = 1.5 + 0.2t^2$
	2	2.0	1	
3	1	0	0	$\lambda^3_{32}(t) = 1.2 + 0.15t$
	2	1.8	0	$\lambda^3_{21}(t) = 2.0 + 0.2t$
	3	4.0	1	

4 Consider a two-component cold standby system with a single repair facility which appears and disappears from the system randomly. The state transition diagram is given in Figure 4.9. In this diagram, States 4, 5, and 6 are down states, where state 6 is a nonregenerative one. The letters a, b, and c denote constant transition rates (exponential distributions of the transition times) for transitions between states, and $G(t)$ is the distribution function of the repair time. Write the cumulative semi-Markov kernel of the system.

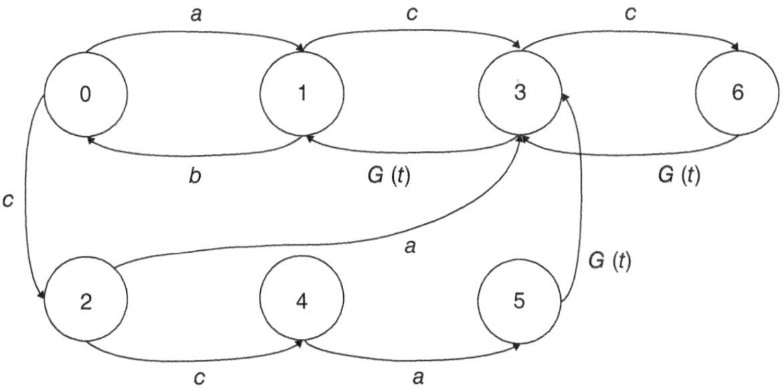

Figure 4.9 State Transition Diagram for a Two-component Cold Standby System with a Single Repair Facility

5 Consider a M-component series system. The random variables J_n, $n=1,2,\dots,M$ denote the type (number) of the component that has failed at the nth-failure. The lifetimes of the components are mutually stochastic independent and exponentially distributed, with parameters λ_k and $k=1,2,\dots,M$. The time $X_n = T_n - T_{n-1}$ is the sum of the lifetime and repair time. The repair times have distribution functions F_k, $k=1,2,\dots,M$, with finite expectations $\mu_k, k=1,2,\dots,M$. The process $(J_n, T_n)_{n\geq 0}$ is a Markov renewal process with cumulative semi-Markov kernel $Q(t)$. Let $\phi(t)$ be the indicator function of the state of the system, i.e. $\phi(t)=1$ means the system is functioning at time t and $\phi(t)=0$ means the system is under repair at time t. Define also the process $(Z_t, t\geq 0)$, which indicates the number of the last failed components before t and put $L_{ij}(t) = \Pr\{Z_t = j \mid \phi(t)=0\}$.

a. Write the Markov renewal equation of $L_{ij}(t)$.
b. If $S(t)$ denotes the number of the hitting times in a renewal process, then the following relationship will exist:

$$\lim_{t\to\infty} \int_0^t \varphi(t-x)\,dS(x) = \frac{1}{\rho}\int_0^\infty \varphi(y)\,dy.$$

The abovementioned equation is called the renewal theorem.
Calculate $\lim_{t\to\infty} L_{ij}(t)$.

References

1 Black, M., Brint, A.T., and Brailsford, J.R. (2005). A semi-Markov approach for modelling asset deterioration. *Journal of the Operational Research Society* 56 (11): 1241–1249.

2 Kim, J. and Makis, V. (2009). Optimal maintenance policy for a multi-state deteriorating system with two types of failures under general repair. *Computers & Industrial Engineering* 57 (1): 298–303.

3 Chryssaphinou, O., Limnios, N., and Malefaki, S. (2011). Multi-state reliability systems under discrete time semi-markovian hypothesis. *IEEE Transactions on Reliability* 60 (1): 80–87.

4 Unwin, S.D., Lowry, P.P., Layton, R.F., Heasler, P.G., and Toloczko, M.B. (2011). Multi-state physics models of aging passive components in probabilistic risk assessment. In: *Proceedings of ANS PSA 2011 International Topical Meeting on Probabilistic Safety Assessment and Analysis*, 1–12.

5 Cloth, L., Jongerden, M.R., and Haverkort, B.R. (2007). Computing battery lifetime distributions. Presented at the 37th Annual IEEE/IFIP International Conference on Dependable Systems and Networks, 2007 (DSN '07), Edinburgh.

6 Liu, Y. and Kapur, K.C. (2008). New model and measurement for reliability of multi-state systems. In: *Handbook of Performability EngineeringLondon* (ed. Krishna B. Misra), Springer.

7 Zio, E. (2009). *Computational Methods for Reliability and Risk Analysis*. World Scientific Publishing Company.

8 Bogachev, V.I., Krylov, N.V., Röckner, M., and Shaposhnikov, S.V. (2015). *Fokker-Planck-Kolmogorov Equations*. American Mathematical Soc.

9 Kexue, L. and Jigen, P. (2011). Laplace transform and fractional differential equations. *Applied Mathematics Letters* 24 (12): 2019–2023.

10 Cocozza-Thivent, C., "Processus de renouvellement markovien, Processus de Markov déterministes par morceaux," Online book available on the webpage: https://hal.archives-ouvertes.fr/hal-01418366/document. (accessed 4 January 2022), 2011.

11 Lin, Y.H., Li, Y.F., and Zio, E. (2015). A Reliability Assessment Framework for Systems With Degradation Dependency by Combining Binary Decision Diagrams and Monte Carlo Simulation. *IEEE Transactions on Systems Man & Cybernetics Systems* 46 (11): 1556–1564.

12 Lewis, E. and Böhm, F. (1984). Monte Carlo simulation of Markov unreliability models. *Nuclear Engineering and Design* 77 (1): 49–62.

13 Cocozza-Thivent, R.E.C. and Mercier, S. (2006). A finite-volume scheme for dynamic reliability models. *IMA Journal of Numerical Analysis* 26: 446–471.

14 Lin, Y.-H., Li, Y.-F., and Zio, E. (2015). Fuzzy reliability assessment of systems with multiple-dependent competing degradation processes. *Fuzzy Systems, IEEE Transactions On* 23 (5): 1428–1438.

5

Monte Carlo Simulation (MCS) for Reliability and Availability Assessment

5.1 Introduction

The Monte Carlo simulation (MCS) may be the only method that can yield solutions to complex multi-dimensional stochastic modeling problems, such as those typically involved in the reliability and availability analysis of real systems.

MCS performs a type of random experiment on a computer, and its development and implementation require domain knowledge in different fields, such as probability and statistics. Its key principle is the repeated random sampling to obtain numerical results. MCS is typically used in three kinds of problems [1]: numerical integration, optimization, and sample generation from a probability distribution, e.g. for uncertainty propagation.

5.2 Random Variable Generation

The foundation of MCS is random variable generation. The foundation of random variable generation is random number generation.

Generally, there are two steps to generate a random variable V from any distribution. The first step is to draw random numbers, $U_1, U_2, \ldots, U_n$, from a uniform distribution in the unit hypercube $(0,1)^n$, and the second step is to return $V = f(U_1, U_2, \ldots, U_n)$ where f is a function from $(0,1)^n$ to $\mathbb{R}^d$. The first step will be discussed in Section 5.2.1; the second step will be discussed in Section 5.2.2.

5.2.1 Random Number Generation

Random number generation is a method to create an infinite set of random numbers, which are independent and identically distributed (iid). The uniform distribution on the interval (0,1) is called a uniform random number generator. On a computer, when a user inputs an initial number, which is called seed, the uniform random number generator

Reliability Analysis, Safety Assessment and Optimization: Methods and Applications in Energy Systems and Other Applications, First Edition. Enrico Zio and Yan-Fu Li.

produces a series of independent uniform random numbers in the interval (0,1). It is the basis of all other random number generators.

Note that the concept of an infinite set of random numbers is only a mathematical abstraction, which is impossible for the contemporary computers to implement. The best one can hope is to generate a sequence of random numbers with statistical properties that are indistinguishable from a sequence of truly random values. In fact, some physical generators based on universal radiation or quantum mechanics might offer such a stable randomness. Nevertheless, most generation methods are based on numerical algorithms implemented in computers. Such algorithms can be described as a tuple (S, f, μ, U, g) where

- S is a finite set of states;
- f is a function from S to S;
- μ is a probability distribution on S;
- U is the output space;
- g is a function from S to U.

The algorithm is made up of four steps:

1) Initialize: Choose the seed S_0 from the distribution μ on S. Set $t = 1$;
2) Transition: Set $S_t = f(S_{t-1})$;
3) Output: Set $U_t = g(S_t)$;
4) Repeat: Set $t = t + 1$ and go back to step 2.

Below are some of the properties of a good random number generator.

- Passing statistical tests: The goal of the random number generator is to generate random numbers, which are indistinguishable from genuine uniform random numbers. So, it is necessary for the random numbers to pass specific statistical tests verifying respective properties.
- Theoretical support: A good generator should be based on a sound mathematical principle and should allow for the analysis of its properties.
- Reproducible: An important property of the generator is to be reproducible, such that it is unnecessary to store the entire sequence of numbers to repeat the outcome. This is important for testing and for comparison to other techniques. Normally, physical generation methods are not reproducible unless the generation progress is recorded.
- Fast and efficient: A good generator should be able to produce numbers in a fast and efficient way and not require much storage in computer memory. Some Monte Carlo methods for optimization or estimation require plenty of random numbers, which cannot be produced by current physical generation methods.
- Large period: The period of a good random number generator (the number of iterations before the sequence of random values returns repeatedly) should be very large, normally of the order of 10^{50}. This is to avoid the repetition of the same sequence, which would introduce dependence in the outcomes.
- Multiple streams: For many applications, running multiple independent random number sequences in parallel is essential.

- Cheap and easy: A good random number generator should be easy to install, implement, and run. Generally, such a random number generator should be portable and used universally.
- Not produce 0 or 1 values: An ideal property of a random number generator is to exclude 0 and 1 from the number sequence, so as to avoid numerical complications in their use for calculation.

5.2.1.1 Linear Recurrences

Linear recurrences are the most common methods to generate pseudorandom numbers. A linear congruential generator generates numbers with the algorithm

$$X_t = \left(\alpha X_{t-1} + c\right) mod\ m, t = 1,2,\ldots,$$

with the state $S_t = X_t \in \{0,\ldots,m-1\}$ and where α (multiplier) and c (increment) are integers. Normally, the outcomes are of the form $U_t = \frac{X_t}{m}$, which gives values in (0,1).

Example 5.1 Lewis, Goodman, and Miller [2] chose $\alpha = 7^5 = 16807$, $c = 0$, and $m = 2^{31} - 1 = 2147483647$. This setting passed lots of the standard statistical tests and has been used successfully in different applications. The method is called minimal standard linear congruential generator (LCG) and used for comparison with other generators.

Though it has some good properties, its period $\left(2^{31} - 1\right)$ is not long enough to meet the requirements of Quasi Monte Carlo methods.

Quasi Monte Carlo methods are MCS methods where the ordinary uniform random points are replaced by quasirandom points. Quasirandom numbers are like random numbers but are present with more regularity, which makes them more suited for numerical evaluation of multi-dimensional integrals. The main types of quasirandom sequences include Halton, Faure, Sobol, and Korobov sequences [3].

5.2.2 Random Variable Generation

The generation of uniform random numbers was introduced in the previous section. This section illustrates how to implement the second step of transforming the random numbers into the values of random variables. Typical methods for generating random variables involve the inverse-transform method, the composition method, the acceptance-rejection method, etc. This section mainly considers the inverse-transform method and the acceptance-rejection method.

5.2.2.1 Inverse-transform Method

Assume $\mathbf{X}$ to be a random variable with cumulative distribution function (cdf) F. The inverse function F^{-1} can be defined as

$$F^{-1}(y) = \inf\{x : F(x) \geq y\}, 0 \leq y \leq 1$$

Take the uniform random number $u \sim \mathrm{U}(0,1)$. The cdf of the inverse transform $F^{-1}(u)$ is given by

$$\mathbb{P}\left(F^{-1}(u) \leq x\right) = \mathbb{P}\left(u \leq F(x)\right) = F(x)$$

Therefore, to generate a random variable X with cdf F, one can generate $u \sim U(0,1)$ and set $X = F^{-1}(u)$.

Generally, the inverse-transform method requires that the cdf F can be presented in a form for which the inverse function F^{-1} can be derived analytically or algorithmically. Thus, the common applicable distributions are exponential distributions, uniform distributions, Cauchy distributions, etc. However, for certain distributions, it is hard to find the inverse transform, which is required to solve

$$F(x) = \int_{-\infty}^{x} f(t)dt = u$$

with respect to x. Even in the situation where F^{-1} exists in an analytical form, the inverse-transform may not be the most efficient method to generate random variables.

Example 5.2 Assume we want to obtain a sample from an exponential probability distribution $f(x) = \lambda e^{-\lambda x}$. We should first generate uniform random numbers $u \sim \mathrm{U}(0,1)$. Then, we can derive the cdf:

$$F(x) = 1 - e^{-\lambda x}$$

Then, we invert to get

$$F^{-1}(x) = -\frac{\ln(1-x)}{\lambda}.$$

From this, we can obtain the realization of the exponential random variable X as $X \sim F^{-1}(u)$. The histograms in Figure 5.1 show the results and a comparison with the true exponential distribution.

5.2.2.2 Acceptance-rejection Method

Let $f(x)$ and $g(x)$ be two probability density functions (pdfs), which satisfy that for some $C \geq 1$, $Cg(x) \geq f(x)$ for all x. Let $X \sim g(x)$ and $u \sim U(0,1)$ be independent. Then, the conditional pdf of X given $u \leq \frac{f(X)}{Cg(X)}$ is $f(x)$. This theorem can be proved [4]. $g(x)$ is the proposal pdf, which is chosen easy to generate random variables from it. The algorithm of the acceptance-rejection method is as follows [4]:

1) Draw X from $g(x)$.
2) Draw U from $U(0,1)$, which is independent of X.

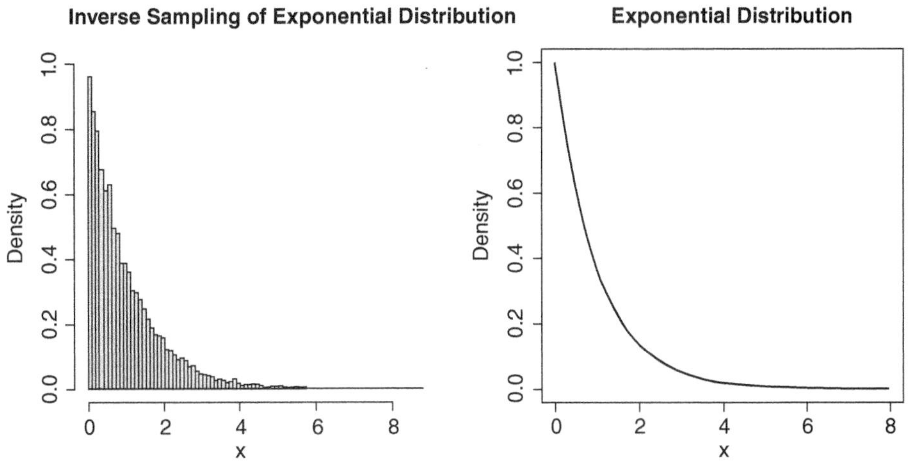

Figure 5.1 comparison between inverse sampling of exponential distribution and exponential distribution.

3) If $U \le \frac{f(X)}{Cg(X)}$, output X; otherwise, return to step 1.

The efficiency of the acceptance-rejection method depends on the probability of acceptance, which is

$$\mathbb{P}\left(U \le \frac{f(X)}{Cg(X)}\right) = \int g(x) \int_0^1 I_{\left\{u \le \frac{f(x)}{Cg(x)}\right\}} dudx = \int \frac{f(x)}{C} dx = \frac{1}{C}$$

Example 5.3 Consider the random variable x with probability distribution (see Figure 5.2)

$$f(x) = \begin{cases} 3x + 0.4 (0 \le x < 0.4) \\ 2.4 - 2x (0.4 \le x \le 1) \end{cases}$$

We can first generate $U_0 \sim \mathrm{U}(0,1)$ where g(x) is the pdf of U_0. As

$$3g(x) \ge f(x)$$

We can generate $U_1 \sim \mathrm{U}(0,1)$, which is independent of U_0. Draw x from the distribution $g(x)$; if $U_1 \le \frac{f(X)}{3g(X)}$, we will accept x as a sample from the distribution $f(x)$. If not, we will reject x.

From Figure 5.2, we see that the distribution obtained from the samples derived by the acceptance-rejection method is similar to the real distribution.

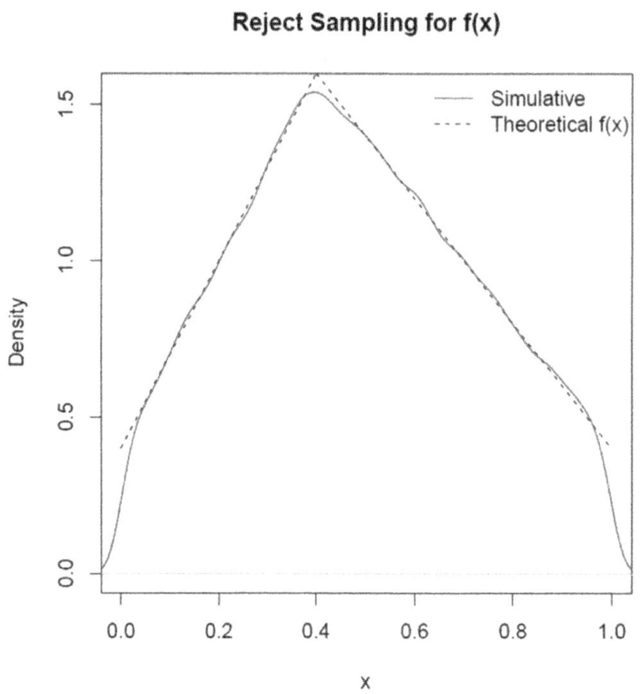

Figure 5.2 Distribution obtained by acceptance-rejection method.

5.2.2.3 Multivariate Random Variable Generation

In this section, we address the problem of generating a vector of values of random variables $\mathbf{X} = (X_1,\ldots,X_n)^T$ with a given joint pdf $f(\mathbf{x})$. When the components $X_1,\ldots, X_n$ are independent, it is easy to solve the problem. Assume the X_i component's marginal pdf is $f_i, i = 1,\ldots,n$; then, $f(\mathbf{x}) = \Pi(f_1,\ldots,f_n)$. To generate the vector $\mathbf{X}$, we can repeatedly draw $X_1,\ldots,X_n$ from their marginal pdfs $f_1,\ldots,f_n$ with the methods referred in the last section:

1) Draw $X_1,\ldots,X_n$ from pdfs $f_1,\ldots,f_n$ independently.
2) Return $\mathbf{X} = (X_1,\ldots,X_n)^{\mathrm{T}}$.

If the components of $\mathbf{X}$ are dependent, we can draw the joint pdf $f(\mathbf{x})$ by exploiting the product rule: of conditional pdfs:

$$f(x) = f(x_1,\ldots,x_n) = f_1(x_1) f_2(x_2 \mid x_1) \ldots f_n(x_n \mid x_1,\ldots,x_{n-1}),$$

where $f_1(x_1)$ is the marginal pdf of X_1 and $f_k(x_k \mid x_1,\ldots,x_{k-1})$ is the conditional pdf of X_k given $X_1 = x_1,\ldots, X_{k-1} = x_{k-1}$. The common procedure is as follows:

1) Generate X_1 with the pdf f_1.
2) For $t = 1{:}n{-}1$, given $X_1 = x_1,\ldots,X_t = x_t$, generate X_{t+1} with the pdf $f_{t+1}\left(x_{t+1|x_1,\ldots x_t}\right)$.
3) Return $\mathbf{X} = (X_1,\ldots X_n)^T$.

We can also adapt the multi-dimensional acceptance-rejection method to generate random vector X.

5.3 Random Process Generation

5.3.1 Markov Chains

As seen in Chapter 4, a Markov chain is a stochastic process $\{X_t,\ t \in T\}$ with a countable index set $T \subset \mathbb{R}$, which has the Markov property

$$(X_{t+s} \mid X_1,\ldots,X_t) = (X_{t+s} \mid X_t).$$

An important property of the Markov chains is that they can be generated sequentially as follows:

1) Generate X_0 from its distribution. Set $t = 0$.
2) Generate X_{t+1} from the conditional distribution of X_{t+1} given X_t.
3) Set $t = t + 1$ and return to step 2.

Normally, the conditional distribution of X_{t+1} given X_t can be specified as follows. $X_{t+1} = g(t, X_t, U_t)$, $t = 0,1,2\ldots$, where g is an easily evaluated function, and U_t is a random variable, which can be easily generated and may depend on X_t and t.

When the Markov chain $\{X_0, X_1,\ldots\}$ has a discrete state space S and is time-homogeneous, its distribution is completely defined by the distribution of X_0 and the matrix of one-step transition probabilities $\mathbb{P} = (p_{ij})$ where

$$p_{ij} = \mathbb{P}(X_{t+1} = j \mid X_t = i), i, j \in S$$

The conditional distribution of X_{t+1} given $X_t = i$ is then the i-th row of $\mathbb{P}$. The generation of a time-homogeneous Markov chain with finite discrete states follows three steps:

1) Generate X_0 from the initial distribution. Set $t = 0$.
2) Generate X_{t+1} from the discrete distribution depending on the X_t-th row of $\mathbb{P}$.
3) Set $t = t + 1$ and return to step 2.

Example 5.4 Thomas has four doors A, B, C, and D in his house. One day, he finds a mouse in his house. The mouse runs around the house from one door to another door. After a period of observation, Thomas finds that the moves of the mouse follow a probability distribution. The probability of the mouse moving from one door to another only depends on the door where the mouse is. Assume the probability distribution can be described by the following transition matrix P. Let X_t be the position of the mouse at time t. Then, $\{X_t\}$ is a time-homogeneous Markov chain with the matrix P:

$$P = \begin{bmatrix} 0 & \frac{1}{2} & \frac{1}{3} & \frac{1}{6} \\ \frac{1}{2} & 0 & \frac{1}{3} & \frac{1}{6} \\ \frac{1}{3} & \frac{1}{3} & 0 & \frac{1}{3} \\ \frac{1}{6} & \frac{1}{2} & \frac{1}{3} & 0 \end{bmatrix}$$

Realization of the Markov chain can be obtained as explained before and the state distribution is plotted in the Figure 5.3.

5.3.2 Markov Jump Processes

Different from the Markov chain, the Markov jump process is a stochastic process $\{X_t, t \in T\}$ with a continuous index set $T \subseteq \mathbb{R}$ and a discrete state space S, which has the Markov property

$$\left(X_{t+s} \mid X_1, \ldots, X_t\right) = \left(X_{t+s} \mid X_t\right).$$

Assume the index set is $T = [0, \infty]$, and the state space is $S = \{1, 2, \ldots\}$.

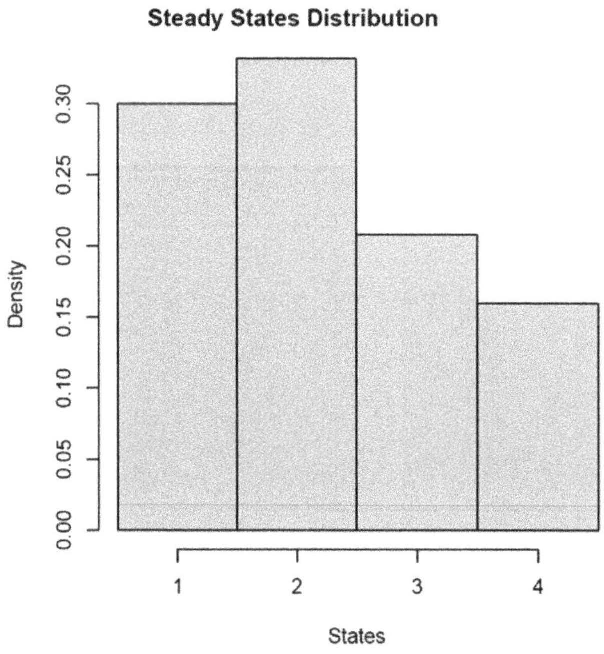

Figure 5.3 State distribution of Markov chain.

A time-homogeneous Markov jump process is usually described by its Q-matrix,

$$Q = \begin{bmatrix} -q_1 & q_{12} & q_{13} & \cdots \\ q_{21} & -q_2 & q_{23} & \cdots \\ q_{31} & q_{32} & -q_3 & \cdots \\ \vdots & \vdots & \vdots & \ddots \end{bmatrix},$$

where q_{ij} is the transition rate from state i to state j:

$$q_{ij} = \lim_{t \to 0} \frac{\mathbb{P}(X_{k+t} = j \mid X_k = i)}{t}, \ \mathbf{i} \in S,$$

and q_i is the holding rate in i:

$$q_i = \lim_{t \to 0} \frac{1 - \mathbb{P}(X_{k+t} = i \mid X_k = i)}{t}, \ i \in S$$

Usually, we assume that $0 \le q_{ij} < \infty$ and that $q_i = \sum_{j \ne i} q_{ij}$, so the sum of each row is 0. An important behavior of such a Markov jump process is as follows. If the process is in a certain state i at time t, it will remain in the same state for an additional $\text{Exp}(q)$-distributed amount of time. Once the process leaves the state i, it will jump to another state j with a probability of $p_{ij} = \frac{q_{ij}}{q_i}$, no matter the history of the process. In particular, the process can be analyzed as a Markov chain. The jump states S_0, S_1,... form a Markov chain with the transition matrix $\text{P} = (P_{ij})$. We can define the holding time as H_1, H_2,... and the jump times as J_1, J_2,..., and the generation procedures are as follows:

1) Set $J_0 = 0$. Generate S_0 from its distribution. Set $X_0 = S_0$ and $n = 0$.
2) Generate $H_{n+1} \sim Exp(q_{S_n})$.
3) Set $J_{n+1} = J_n + H_{n+1}$.
4) Set $X_t = S_n$ for $J_n \le t < J_{n+1}$.
5) Generate S_{n+1} from the distribution related to the S_n-th row of P. Set $n = n + 1$ and return to step 2.

Example 5.5 Assume there are two babies and one babysitter. Both babies have exponentially distributed waking times and times for babysitter to make them sleep. The waking and babysitting rates are respectively a_1, b_1, a_2, b_2. The babysitter can only babysit one baby at a time. If two babies are awake, the babysitter will keep babysitting the baby who is first awake. The system can be seen as a Markov jump process with five states: 1 (both babies are asleep); 2 (baby 1 wakes and baby 2 is asleep); 3 (baby 2 wakes and baby 1 is asleep); 4 (both babies wake and baby 1 wakes first); 5 (both babies wake and baby 2 wakes first). The transition matrix is as follows. Assume $a_1 = 1$, $a_2 = 2$, $b_1 = 3$, $b_2 = 4$.

$$Q = \begin{pmatrix} -(a_1+a_2) & a_1 & a_2 & 0 & 0 \\ b_1 & -(b_1+a_2) & 0 & a_2 & 0 \\ b_2 & 0 & -(b_2+a_1) & 0 & a_1 \\ 0 & 0 & b_1 & -b_1 & 0 \\ 0 & b_2 & 0 & 0 & -b_2 \end{pmatrix}$$

We can see the Markov jump process from Figure 5.4.

When it comes to the nonhomogeneous case, the algorithm is similar, but the rates appearing in matrix Q depend on time T. Replace q_{ij} with $q_{ij}(t)$ and let $q_i(t) = \sum_{j \neq i} q_{ij}(t)$. The process jumps from a state to another depending on a time-nonhomogeneous Markov chain and stays some time in each state. Assume at a certain time, T_n, the process jumps to state $S_n = \mathbf{i}$. Let H_{n+1} denote the holding time in state i. Then,

$$\begin{aligned} q_i(t) &= \lim_{h \to 0} \frac{\mathbb{P}(t - T_n < H_{n+1} < t + h - T_n \mid H_{n+1} > t - T_n)}{h} \\ &= \lim_{h \to 0} \frac{F(t+h-T_n) - F(t-T_n)}{(1 - F(t-T_n))h} = \frac{\mathbf{f}(\mathbf{t} - T_n)}{1 - F(t - T_n)} \\ &= -\frac{\mathbf{d}}{\mathbf{dt}} \ln(1 - F(t - T_n)), \end{aligned}$$

where $F(t)$ is the cdf of H_{n+1}, and $f(t)$ its pdf. We can get $F(t)$ by using

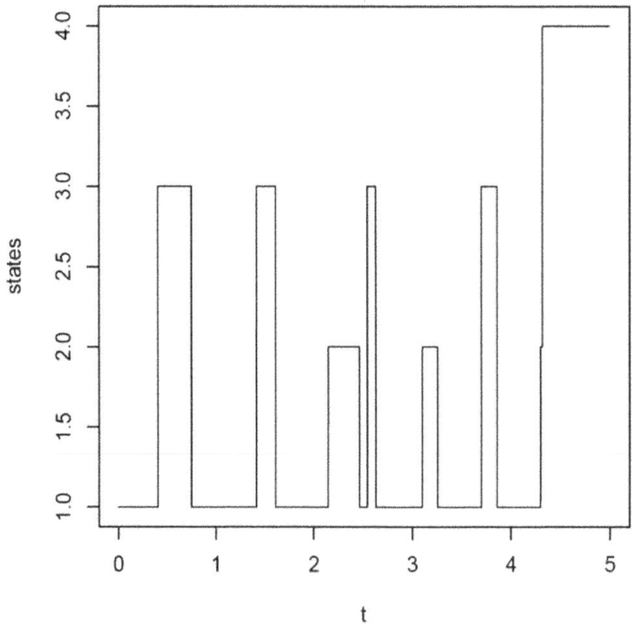

Figure 5.4 Markov jump process.

$$F(t)=\mathbb{P}(H_{n+1}\leq t)=1-e^{-\int_{T_n}^{T_n+t} q_i(s)ds}, t\geq 0$$

At time $T_{n+1}=T_n+H_{n+1}$, the process jumps to state j with probability $\frac{q_{ij}(T_{n+1})}{q_i(T_{n+1})}, j\in S$. So, we get the following algorithm:

1) Set $T_0=0$. Generate S_0 according to its distribution. Set $X_0=Y_0$ and $n=0$.
2) Generate H_{n+1} from the cdf given above.
3) Set $T_{n+1}=T_n+H_{n+1}$.
4) Set $X_t=S_n$ for $T_n\leq t<T_{n+1}$.
5) Generate S_{n+1} from the distribution $\left\{\frac{q_{S_n,S}(T_{n+1})}{q_{S_n}(T_{n+1})}, y\in S\right\}$. *Set n = n + 1* and return to step 2.

5.4 Markov Chain Monte Carlo (MCMC)

Markov Chain Monte Carlo (MCMC) is a key method for sampling from a given distribution. By means of generating a Markov chain which has the desired distribution as its limiting distribution, we can get a sample of the desired distribution by observing the chain after a few steps. In this section, we will describe three most prominent MCMC algorithms:

1) The Metropolis-Hastings (M-H)algorithm, i.e. the independence sampler and random walk sampler [5];
2) The Gibbs sampler, which is very useful in Bayesian analysis [5];
3) Multiple-try Metropolis-Hastings method where different algorithms are combined [5].

5.4.1 Metropolis-Hastings (M-H) Algorithm

The M-H algorithm is similar to the acceptance-rejection algorithm to some degree. Let $f(x)$ be a function, which is proportional to the desired probability distribution $p(x)$:

1) Initialize with some X_0 as the first sample, and select an arbitrary probability $q(y\mid x)$ as a proposal or instrumental density, which is used to generate the next sample y given x.
2) Generate y with the distribution $q(y\mid x)$ given X_0.
3) Calculate the acceptance ratio $\alpha(x,y)=\min\left\{\frac{f(y)q(x\mid y)}{f(x)q(y\mid x)},1\right\}$.
4) Generate a uniform random number $u\sim \mathrm{U}(0,1)$. If $u\leq\alpha(X_t,y)$, then set $X_{t+1}=y$. If not, then set $X_{t+1}=X_t$.

The probability $\alpha(x,y)$ is called the acceptance probability. The algorithm proceeds by randomly accepting the moves or remaining in place. We can see that α can represent how probable the new sample is, given the current one. So, when α is large enough ($\alpha\geq 1$), one accepts the new sample; otherwise, there is some possibility that it remains in place. Therefore, there is a tendency to stay in high-density regions of $p(x)$.

We, thus, get the so-called M-H Markov chain, $X_0, X_1, \ldots X_T$, with X_T approximately distributed as $f(x)$ for large T.

Example 5.6 Assume we want to sample from a distribution $p(x)$. We only know function $f(x)$ where $f(x) = \frac{p(x)}{Z}$ and $f(x) = x^2 e^{-x}$.

We can sample from $p(x)$ as just explained.

We can see from Figure 5.5 above that the result obtained is close to the real distribution.

5.4.2 Gibbs Sampler

A Gibbs sampler can be seen as a special case of the M-H algorithm for generating n-dimensional random vectors [6]. One of the most distinguishing features of the Gibbs sampler is that the corresponding Markov chain is constructed from a set of conditional distributions in either a deterministic or random form. Gibbs sampling is useful when the joint distribution is unknown or difficult to sample from directly, but the conditional distribution of each variable is known and easy to sample from.

Assume that we want to sample a random vector $\mathbf{X} = (X_1, \ldots, X_n)$ according to a target pdf $f(x)$. Suppose that $f(x_i \mid x_1, \ldots, x_{i-1}, x_{i+1}, \ldots, x_n)$ represents the conditional pdf of the ith component of the vector X. The Gibbs sampler algorithm is as follows:

1) Initialize with a state X_0. Set $t = 0$.
2) For a given $\mathbf{X}_t$, draw $\mathbf{Y} = (Y_1, \ldots, Y_n)$ as follows:
   - Generate Y_1 from the conditional distribution $f(x_1 \mid x_{t,2}, \ldots, x_{t,n})$.
   - Generate Y_i from $f(x_i \mid Y_1, \ldots, Y_{i-1}, x_{t,i+1}, \ldots, x_{t,n})$.
   - Generate Y_n from $f(x_n \mid Y_1, \ldots, Y_{n-1})$.
3) Set $\mathbf{X}_{t+1} = \mathbf{Y}$.

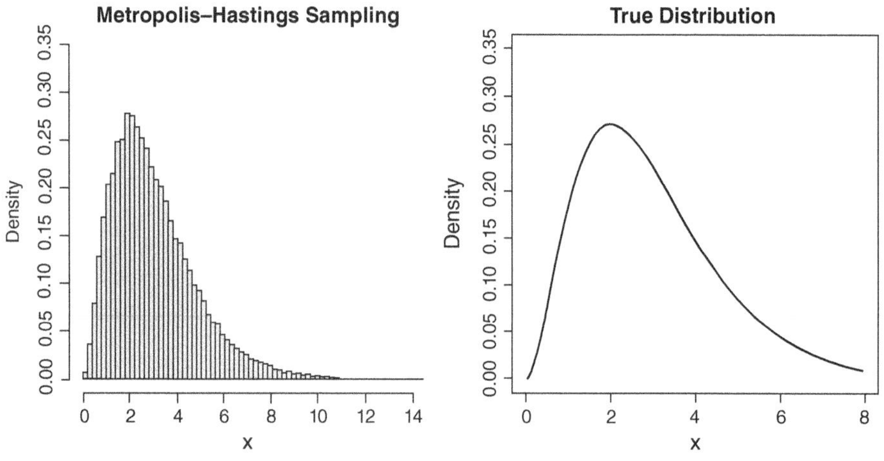

Figure 5.5 M-H sampling.

Example 5.7 Assume that $\mathbf{X}^{(i)} = \left(x^{(i)}, y^{(i)}\right)$ is a vector whose components follow a bivariate normal distribution $\boldsymbol{X} \sim \boldsymbol{N}(\mathbf{0}, \Sigma)$, the standard deviations of x and y are 1, and the correlation coefficient between x and y is $r = 0.98$. We can sample $\mathbf{X}$ with Gibbs sampler, with the results shown in Figure 5.6.

5.4.3 Multiple-try Metropolis-Hastings (M-H) Method

The multiple-try M-H algorithm is an extension of the M-H algorithm, which can accelerate the sampling by making the sampling step size larger and the acceptance rate higher.

In M-H method, we often use the normal distribution as the proposal distribution i.e. $Q\left(x', x^t\right) = N\left(x^t, \sigma^2 I\right)$. However, it is difficult to determine the value of σ^2 in $N\left(x^t, \sigma^2\right)$. Although the method is fundamental to converge to the limiting distribution, with a finite sample size in practice, the progress can be slow. If σ^2 is very large, most steps of the sampling will be rejected; if σ^2 is very small, most steps will be accepted, and the Markov chain will be close to a random walk through the probability space.

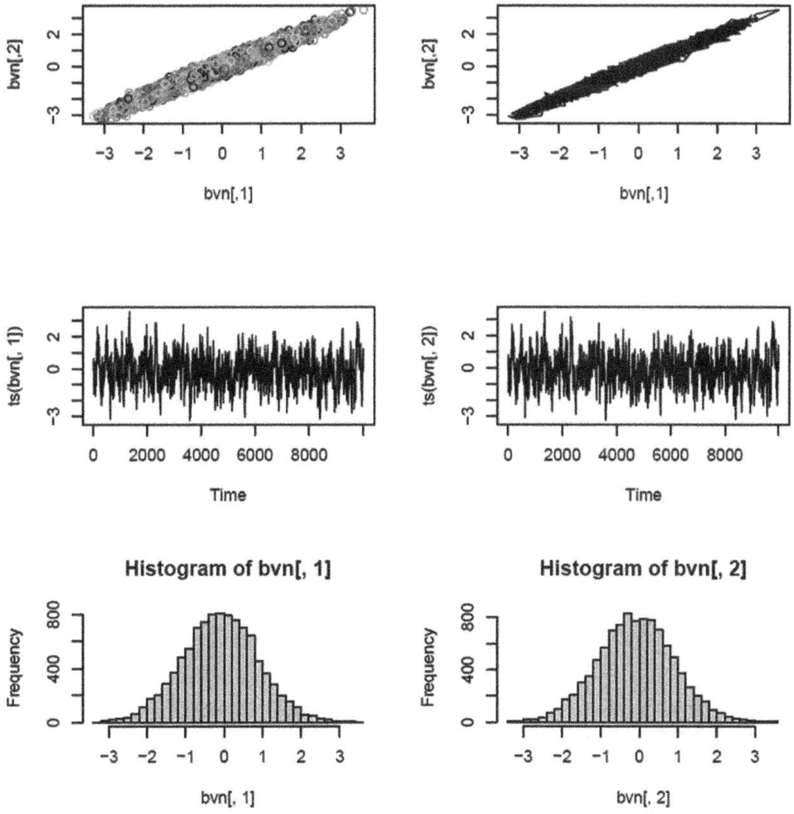

Figure 5.6 Sampling with Gibbs sampler.

Especially when the dimensionality of **x** is high, it is difficult to get the appropriate acceptance rate and step size at the same time.

Multiple-try Metropolis algorithm can resolve this issue. Its procedures are shown as follows. Assume $q(y|x)$ is an arbitrary symmetric proposal function. Initialize the sequence with some X_0 which satisfies $f(X_0) > 0$. Let M be the dimension parameter.

1) Set $t = 0$.
2) Draw proposals $Y_1, \ldots, Y_M \overset{iid}{\sim} q(y|X_t)$.
3) Draw a random index J from the set $\{1, \ldots, M\}$, such that

$$\mathbb{P}(J = j) = \frac{f(Y_j)}{f(Y_1) + \ldots + f(Y_M)}, j = 1, \ldots, M.$$

4) Draw proposals $Z_1, \ldots, Z_{M-1} \sim q(z|Y_J)$ and set $Z_M = X_t$ given J.
5) Set $\alpha(X_t, Y_J) = \min\left\{\frac{f(Y_1) + \ldots + f(Y_M)}{f(Z_1) + \ldots + f(Z_M)}, 1\right\}$.

   Draw $u \sim \mathrm{U}(0,1)$. If $u \le \alpha(X_t, Y_J)$, $X_{t+1} = Y_J$; if not, then $X_{t+1} = X_t$.
6) Set $t = t + 1$, return to step 2.

Example 5.8 Assume we want to obtain a sample from a distribution *p(x)*. We only know function $f(x)$ where $f(x) = \frac{p(x)}{Z}$.

$$f(x) = x^2 e^{-x}$$

We can sample from $p(x)$ and the results are shown in Figure 5.7. Compared with Example 5.6 where M-H is applied, it converges faster with multiple-try M-H algorithm.

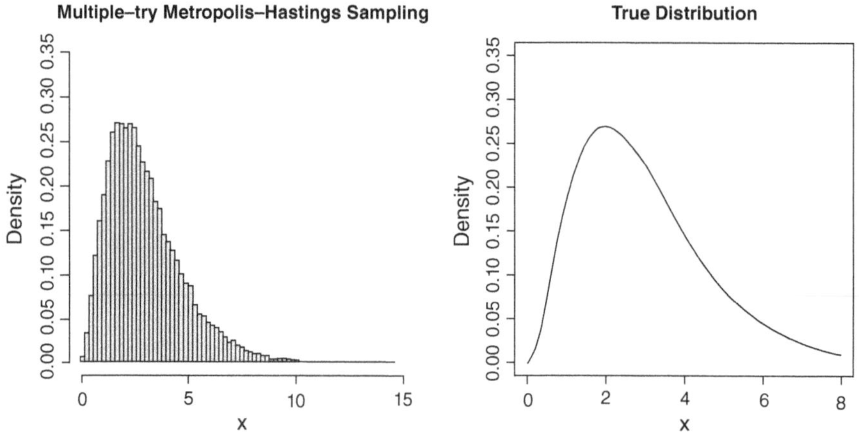

Figure 5.7 Sampling with Multiple-try Metropolis-Hastings (M-H) Method.

5.5 Rare-Event Simulation

Components and systems failure events are typically rare. Let us refer to failure event A of a generic component, whose probability $\alpha = P(A) \ll 1$. The normal, crude Monte Carlo method estimates α by the proportion b of times in which the event A occurs over n repeated independent trials.

$$b = \frac{1}{n}\sum_{j=1}^{n} B_j,$$

where $B_1, B_2, \ldots, B_n$ are binary numbers that indicate the realizations of event A, i.e. $B_i = 1$ means occurrence and $B_i = 0$ overwise. According to the Central Limit Theorem (CLT) for binomial distributions [7], we have

$$\lim_{n\to\infty} \frac{\sqrt{n}\left(b-\alpha\right)}{\sqrt{\alpha\left(1-\alpha\right)}} = N\left(0,1\right)$$

and

$$b \approx \alpha + \sqrt{\frac{\alpha\left(1-\alpha\right)}{n}} N\left(0,1\right)$$

for $n \gg \frac{1}{\alpha}$. The absolute error is

$$b - \alpha \approx \sqrt{\frac{\alpha\left(1-\alpha\right)}{n}} N\left(0,1\right)$$

and the relative error is

$$\frac{b}{\alpha} - 1 \approx \sqrt{\frac{1-\alpha}{\alpha n}} N\left(0,1\right)$$

Therefore, crude Monte Carlo requires that

$$n \gg \frac{1}{\alpha}$$

for the relative error to be small. This makes rare-event simulation expensive in many cases.

5.5.1 Importance Sampling

One of the most widely used methods for rare-event sampling is importance sampling. Let us consider the problem of evaluating $E\left[f\left(x\right)\right], x \sim p$, which is

$$E\left[f\left(x\right)\right] = \int_x f\left(x\right) p\left(x\right) dx$$

To estimate this quantity using MCS, we would need to generate samples from the probability distribution $p(x)$. To do so more efficiently, we can introduce another distribution $q(x)$, from which we can simply draw samples:

$$\int_{x} f(x)p(x)dx = \int_{x} f(x)\frac{p(x)}{q(x)}q(x)dx$$
$$= \int_{X} g(x)q(x)dx$$

where $g(x) = f(x)\frac{p(x)}{q(x)} = f(x)w(x)$.

The original problem has become equivalent to evaluating $E(g(x))$ with the probability distribution $q(x)$ and where $w(x) = \frac{p(x)}{q(x)}$ is called importance weight. When we adopt the importance sampling for rare-event sampling, we can choose $q(x)$ to control the variance of the sampling.

Example 5.9 Consider a function $f(x) = exp(-2|x-6|)$. Assume we want to get $E(f(X))$ where $X \sim U(1,11)$. That is, we want to calculate the integral

$$\int_{1}^{11} \exp(-2|x-6|)dx$$

The normal way to solve the problem is to first generate samples $X \sim U(1,11)$ and then calculate the mean of $10 \cdot f(X)$. The true mean is about 1. In one run, we get 0.9930885.

The function $f(x)$ in this case is peaked at 6 and decreases quickly elsewhere; therefore, by using the uniform distribution, many of the samples contribute little to the expectation. If we use a Gaussian function with a peak at 6 and a small variance, we get greater precision:

$$\int_{1}^{11} \exp(-2|x-6|)\frac{1}{\frac{1}{\sqrt{2\pi}}e^{-\frac{(x-6)^2}{2}}}\frac{1}{\sqrt{2\pi}}e^{-\frac{(x-6)^2}{2}}dx$$

The result is 0.9996468, which is closer to the true value.

5.5.2 Repetitive Simulation Trials after Reaching Thresholds (RESTART)

The Repetitive Simulation Trials After Reaching Thresholds (RESTART) method is an alternative to the crude Monte Carlo computation of rare-event probabilities. The core concept of the RESTART method is that, given a rare event A whose probability must be estimated, an event C is defined so that $C \supset A$ and $1 \gg P(C) \gg P(A)$. The probability of occurrence of event A can be written as [8]

$$P(A) = P(C) \cdot P(A \mid C).$$

In a simulation, $P(C)$ is usually easier to estimate than $P(A \mid C)$, since P(C) is estimated from the whole simulation and P(A|C) is estimated from the small portion of simulations in which C occurs.

The RESTART can be enhanced by defining multiple thresholds C_i that satisfy $C_1 \supset C_2 \supset \ldots \supset C_M \supset A$. The efficiency of the method is greatly improved with multiple thresholds [9].

Let us consider a Markov process $X(t)$. S is the state space of $X(t)$. To make the RESTART method work, we need a function $\phi : S \to \mathbb{R}$, which is called the importance function. Thresholds $T_i (1 \le i \le M)$ of ϕ are defined, so that each set C_i is associated with $\phi \ge T_i$.

When an event occurs where the system is in set A or C_i, the event is called event A or event C_i. Other two kinds of events B_i and D_i are defined as follows:

B_i: instant of the transition from other states to C_i
D_i: instant of the transition from C_i to other states

The RESTART method works as follows:

1) When an event B_i occurs, the system state is saved;
2) When an event D_i occurs, the system state at last event B_i is restored and simulation is conducted again beginning with B_i until it reaches D_i;
3) The process mentioned above is repeated R_i times, which is the number of retrials for threshold i. The starting event of each trial is always the same B_i, while the ending events are different, D_{i1}, D_{i2} ,..., D_{iRi};
4) During one trial of level i, an event B_{i+1} may occur and R_{i+1} trials of level $i+1$ would be made before the trial of level i finishes;
5) When event D_{iR} occurs, simulation continues in the usual way (do not need to start with B_i).

The statistics should be modified accordingly to calculate the probability of event A. The way to modify the statistics is presented as follows. If the estimator of the probability of the rare event in crude simulation is:

$$P = \frac{N_A}{N}$$

where N_A is the number of events A occurred in the simulation and N is the total number of simulated event, then, the estimator with RESTART should be:

$$\hat{P} = \frac{N_A}{N \cdot \prod_{i=1}^{M} R_i}$$

where N_A includes all the events A occurred in all retrials, while N only includes the events simulated in the first trial of each set of retrials.

5.6 Exercises

1. Sample by inverse sampling from an exponential probability distribution $p(x) = \lambda e^{-\lambda x}$ where $\lambda = 2$.

2. Assume there are two babies and one babysitter. Both babies have exponentially distributed waking times and times for the babysitter to make them sleep. The waking and babysitting rates are respectively a_1, b_1, a_2, b_2. The babysitter can only babysit one baby at a time. If both babies are awake, the babysitter will keep babysitting the baby who is first awake. The system can be seen as a Markov jump process with five states: 1 (both babies are asleep); 2 (baby 1 wakes and baby 2 is asleep); 3 (baby 2 wakes and baby 1 is asleep); 4 (both babies wake and baby 1 wakes first); 5 (both babies wake and baby 2 wakes first). The transition matrix is as follows. Assume $a_1 = 2\ a_2 = 3,\ b_1 = 4,\ b_2 = 5$.

$$Q = \begin{pmatrix} -(a_1 + a_2) & a_1 & a_2 & 0 & 0 \\ b_1 & -(b_1 + a_2) & 0 & a_2 & 0 \\ b_2 & 0 & -(b_2 + a_1) & 0 & a_1 \\ 0 & 0 & b_1 & -b & 0 \\ 0 & b_2 & 0 & 0 & -b_2 \end{pmatrix}$$

Sample states with the Markov jump process method ($t \leq 10$).

3. Assume we want to sample from a distribution p(x). We only know function f(x) where $f(x) = \frac{p(x)}{Z}$.

$$f(x) = e^{-2x}$$

Sample from $f(x)$ with M-H algorithm.

4. Estimate $P(X \geq 4)$ where $X \sim N(0,1)$ with importance sampling. (For example, you can use $g(X) = N(5,1)$ as a proposal distribution).

Appendix

R Code for the examples presented in the chapter:

```
#Example 5.2
cdf <- function(f, lower_bound, upper_bound)
{
  if(lower_bound < -1000) lower_bound <- -1000                    # Trim
large negatives
  if(upper_bound > 1000) upper_bound <- 1000                      # Trim
large positive
  x <- seq(lower_bound, upper_bound, length.out = 10000) # Finely
divide x axis
  delta <- mean(diff(x))                                           # Get
delta x (i.e. dx)
  mid_x <- (x[-1] + x[-length(x)])/2                               # Get
the mid point of each slice
  result <- cumsum(delta * f(mid_x))                              # sum f(x)
dx
```

```
  result <- result / max(result)                                    # normal-
ize
  list(x = mid_x, cdf = result)                                     # return
both x and f(x) in list
}
inv_sample <- function(f, n = 1, lower_bound = -1000, upper_bound
= 1000)
{
  CDF <- cdf(f, lower_bound, upper_bound)
  samples <- runif(n)
  sapply(samples, function(s) CDF$x[which.min(abs(s - CDF$cdf))])
}
hist(inv_sample(dexp, 10000, 0, 100), breaks=100,  freq=FALSE,ma
in="Inverse Sampling of Exponential Distribution",xlab="X",ylim
=c(0,1))
par(new=TRUE)
plot(dexp,0,8,main="Exponential
Distribution",xlab="X",ylab="Density",ylim=c(0,1))

#Example 5.3
#f(x) function
fx <- function(x){
  if(x<=0.4) y=3*x+0.4
  else y=2.4-2*x
  return(y)
}
fx1 <- function(x){
  return(3*x+0.4)
}
fx2 <- function(x){
  return(2.4-2*x)
}
accept <- function() {
  while (T) {
    x <- runif(1)   # sample from g~U(0,1)
    u <- runif(1)
    if (u < fx(x)/3)   # Whether accept x
      break
    failtime <<- failtime + 1   # record failure times
  }
  x
}
samplex <-function(n){
  set.seed(123)
  replicate(n,accept())
}
n = 100000
failtime=0  # record failure times
res <- samplex(n)
failrate <- failtime/(failtime+n)
```

```
plot(density(res), xlim=c(0, 1), col="red", xlab="x",
     main="Reject Sampling for f(x)")
curve(fx1, 0, 0.4, col="blue", add=T, lty=2)
curve(fx2, 0.4, 1, col="blue", add=T, lty=2)
legend("topright", legend=c("Simulative", "Theoretical f(x)"),
       col=c("red", "blue"), lty=c(1,2), bty="n")

#example 5-4
# simulate discrete Markov chains with transition matrix P
sim.markov <- function( P, iterations=50) {
  # number of states
  num.states <- nrow(P)
  # stores the states
  states  <- numeric(iterations)
  # initialize variable for first state
  states[1]    <- 1
  for(t in 2:iterations) {
    # probability vector to simulate next state
    p  <- P[states[t-1], ]
    ## draw from multinomial and determine state
    states[t] <-  which(rmultinom(1, 1, p) == 1)
  }
  return(states)
}
P <- t(matrix(c( 0,  1/2, 1/3,1/6,
                1/2,    0, 1/3,1/6,
                1/3,  1/3,   0,1/3,
                1/6,  1/2, 1/3, 0), nrow=4, ncol=4))
num.chains     <- 5
num.iterations <- 50
chain.states <- matrix(NA, ncol=num.chains, nrow=num.iterations)
for(c in seq_len(num.chains)){
  chain.states[,c] <- sim.markov(P)
}
matplot(chain.states, type='l', lty=1, col=1:5, ylim=c(0,4),
ylab='state', xlab='time')
abline(h=1, lty=3)
abline(h=3, lty=3)
count.num = array(table(chain.states))
hist(chain.states,breaks = c(0.5, 1.5, 2.5, 3.5,4.5),freq=FALSE,ma
in="Steady States Distribution",xlab="States")

#Example 5-5
sim.cont.markov <- function(Q, t=5,dt=0.001) {
    # number of states
    num.states <- nrow(Q)
    # probability matrix
    P  <- Q
    diag(P) = rep(0,dim(P)[1]);
    P  <- P/apply(P,1,sum)
```

```
    # stores the states
    iterations = t/dt
    states  <- numeric(iterations)
    # initialize variable for first state
    states[1]    <- 1
    for(t in 2:iterations) {
      # probability vector to simulate next state
      m  <- states[t-1]
      p  <- P[m,]
      set.seed(1)
      ran.num <- runif(1)
      if(ran.num<dt*(-Q[m,m])){
        states[t] <-  which(rmultinom(1, 1, p) == 1)
      }
      else{
        states[t] <- states[t-1]
      }
    }
    return(states)
  }
  a1=1
  a2=2
  b1=3
  b2=4
  Q <- t(matrix(c(-(a1+a2), a1      , a2       ,  0, 0,
                b1      , -(b1+a2), 0        , a2, 0,
                b2      ,  0      , -(b2+a1),  0,a1,
                0       ,  0      , b1       ,-b1, 0,
                0       ,  b2     , 0        ,   0, -b2),
nrow=5,ncol=5))
  t <- seq(0,5,0.001)[-length(t)]
  states <- sim.cont.markov(Q)
  plot(t,states,type="l")

#Example 5-6
  fx = function(x){
    if(x<0){
      return(0)}
    else {
      return( x*x*exp(-x))
    }
  }
  fx2 =function(x){
    return(fx(x)/2)
  }
  x = rep(0,50000)
  x[1] = 1     #starting value
  for(i in 2:50000){
    currentx = x[i-1]
    newx = currentx + rnorm(1,mean=0,sd=1)
```

```
    A = fx(newx)/fx(currentx)
    if(runif(1)<A){
      x[i] = newx       # accept move with probably min(1,A)
    } else {
      x[i] = currentx         # otherwise "reject" move, and stay
where we are
    }
  }
  hist(x,breaks=100,  freq=FALSE,main="Metropolis-Hastings Sampling",
xlab="X",ylim=c(0,0.35))
  plot(fx2,0,8,main="True Distribution",xlab="X",ylab="Density",yl
im=c(0,0.35))

#Example 5-7
  gibbs<-function (n, r)
  {
    mat <- matrix(ncol = 2, nrow = n)
    x <- 0
    y <- 0
    mat[1, ] <- c(x, y)
    for (i in 2:n) {
      x <- rnorm(1, r * y, sqrt(1 - r^2))
      y <- rnorm(1, r * x, sqrt(1 - r^2))
      mat[i, ] <- c(x, y)
    }
    mat
  }
  bvn<-gibbs(10000,0.98)

  par(mfrow=c(3,2))
  plot(bvn,col=1:10000)
  plot(bvn,type="l")
  plot(ts(bvn[,1]))
  plot(ts(bvn[,2]))
  hist(bvn[,1],40)
  hist(bvn[,2],40)
  par(mfrow=c(1,1))

#Example 5-8
  fx = function(x){
    if(x<0){
      return(0)}
    else {
      return( x*x*exp(-x))
    }
  }
  fx2 =function(x){
    return(fx(x)/2)
  }
  n = 10
```

```
x = rep(0,50000)
y = rep(0,10)
z = rep(0,10)
x[1] = 1     #starting value
for(i in 2:50000){
  currentx = x[i-1]
  for(j in 1:10){
    y[j]=currentx + rnorm(1,mean=0,sd=1)
  }
  sumy <- 0
  for(k in 1:10){
    sumy <- sumy + fx(y[k])
  }
  ynorm = rep(0,10)
  for(k in 1:10){
    ynorm[k]=fx(y[k])/sumy
  }
  itemy <-  which(rmultinom(1, 1, ynorm) == 1)
  for(j in 1:9){
    z[j] <- y[itemy]+rnorm(1,mean=0,sd=1)
  }
  z[10] <- currentx
  sumz <- 0
  for(k in 1:10){
    sumz <- sumz + fx(z[k])
  }
  alphay <- sumy/sumz
  if(runif(1) < alphay){
    x[i]=y[itemy]
  }
    else{
    x[i]=currentx
  }
}
hist(x,breaks=100,  freq=FALSE,main="Multiple-try Metropolis-
Hastings Sampling",xlab="X",ylim=c(0,0.35))
plot(fx2,0,8,main="True Distribution",xlab="X",ylab="Density",yl
im=c(0,0.35))

#Example 5-9
fx <- function(x){
    return(exp(-2*abs(x-6)))
}
n=10000
x <- array(runif(n,1,11))
x.f <- apply(x,1,fx)
10*sum(x.f)/n

fx2 <- function(x){
  return(exp(-2*abs(x-6))/(1/sqrt(2*pi)*exp(-((x-6)**2)/2)))
```

```
  }
  y <- array(rnorm(n,6,1))
  y.f <- apply(y,1,fx2)
  10*sum(y.f)/n
  if(lower_bound < -1000) lower_bound <- -1000    # Trim large nega-
tives
  if(upper_bound > 1000) upper_bound <- 1000    # Trim large
positive
  x <- seq(lower_bound, upper_bound, length.out = 10000) # Finely
divide x axis
  delta <- mean(diff(x))                        # Get delta x (i.e.
dx)
  mid_x <- (x[-1] + x[-length(x)])/2                      # Get
the mid point of each slice
  result <- cumsum(delta * f(mid_x))                     # sum f(x)
dx
  result <- result / max(result)                          # normal-
ize
  list(x = mid_x, cdf = result)                        # return both
x and f(x) in list
}
inv_sample <- function(f, n = 1, lower_bound = -1000, upper_bound
= 1000)
{
  CDF <- cdf(f, lower_bound, upper_bound)
  samples <- runif(n)
  sapply(samples, function(s) CDF$x[which.min(abs(s - CDF$cdf))])
}
hist(inv_sample(dexp, 10000, 0, 100), breaks=100,  freq=FALSE,main
="Inverse Sampling of Exponential Distribution",xlab="X",ylim
=c(0,1))
par(new=TRUE)
plot(dexp,0,8,main="Exponential Distribution",xlab="X",ylab="
Density",ylim=c(0,1))

#Example 5.3
#f(x) function
fx <- function(x){
  if(x<=0.4) y=3*x+0.4
  else y=2.4-2*x
  return(y)
}
fx1 <- function(x){
  return(3*x+0.4)
}
fx2 <- function(x){
  return(2.4-2*x)
}
accept <- function() {
  while (T) {
```

```
    x <- runif(1)   # sample from g~U(0,1)
    u <- runif(1)
    if (u < fx(x)/3)   # Whether accept x
      break
    failtime <<- failtime + 1   # record failure times
  }
  x
}

samplex <-function(n){
  set.seed(123)
  replicate(n,accept())
}
n = 100000
failtime=0  # record failure times
res <- samplex(n)
failrate <- failtime/(failtime+n)
plot(density(res), xlim=c(0, 1), col="red", xlab="x",
     main="Reject Sampling for f(x)")
curve(fx1, 0, 0.4, col="blue", add=T, lty=2)
curve(fx2, 0.4, 1, col="blue", add=T, lty=2)
legend("topright", legend=c("Simulative", "Theoretical f(x)"),
       col=c("red", "blue"), lty=c(1,2), bty="n")

#example 5-4
# simulate discrete Markov chains with transition matrix P
sim.markov <- function( P, iterations=50) {
  # number of states
  num.states <- nrow(P)
  # stores the states
  states  <- numeric(iterations)
  # initialize variable for first state
  states[1]    <- 1
  for(t in 2:iterations) {
    # probability vector to simulate next state
    p  <- P[states[t-1], ]
    ## draw from multinomial and determine state
    states[t] <-  which(rmultinom(1, 1, p) == 1)
  }
  return(states)
}
P <- t(matrix(c( 0,  1/2, 1/3,1/6,
                1/2,    0, 1/3,1/6,
                1/3,  1/3,   0,1/3,
                1/6,  1/2, 1/3, 0), nrow=4, ncol=4))
num.chains     <- 5
num.iterations <- 50
chain.states <- matrix(NA, ncol=num.chains, nrow=num.iterations)
for(c in seq_len(num.chains)){
```

```
  chain.states[,c] <- sim.markov(P)
}
matplot(chain.states, type='l', lty=1, col=1:5, ylim=c(0,4),
ylab='state', xlab='time')
abline(h=1, lty=3)
abline(h=3, lty=3)
count.num = array(table(chain.states))
hist(chain.states,breaks = c(0.5, 1.5, 2.5, 3.5,4.5),freq=FALSE,ma
in="Steady States Distribution",xlab="States")

  #Example 5-5
  sim.cont.markov <- function(Q, t=5,dt=0.001) {
    # number of states
    num.states <- nrow(Q)
    # probability matrix
    P  <- Q
    diag(P) = rep(0,dim(P)[1]);
    P  <- P/apply(P,1,sum)
    # stores the states
    iterations = t/dt
    states   <- numeric(iterations)
    # initialize variable for first state
    states[1]     <- 1
    for(t in 2:iterations) {
      # probability vector to simulate next state
      m  <- states[t-1]
      p  <- P[m,]
      set.seed(1)
      ran.num <- runif(1)
      if(ran.num<dt*(-Q[m,m])){
        states[t] <-  which(rmultinom(1, 1, p) == 1)
      }
      else{
        states[t] <- states[t-1]
      }
    }
    return(states)
  }
  a1=1
  a2=2
  b1=3
  b2=4
  Q <- t(matrix(c(-(a1+a2), a1      , a2       ,  0, 0,
                  b1      , -(b1+a2), 0        , a2, 0,
                  b2      , 0       , -(b2+a1),  0,a1,
                  0       , 0       , b1       ,-b1,0,
                  0       , b2      , 0        ,  0,
-b2),nrow=5,ncol=5))
  t <- seq(0,5,0.001)[-length(t)]
```

```
  states <- sim.cont.markov(Q)
  plot(t,states,type="l")

#Example 5-6
  fx = function(x){
    if(x<0){
      return(0)}
    else {
      return( x*x*exp(-x))
    }
  }
  fx2 =function(x){
    return(fx(x)/2)
  }
  x = rep(0,50000)
  x[1] = 1     #starting value
  for(i in 2:50000){
    currentx = x[i-1]
    newx = currentx + rnorm(1,mean=0,sd=1)
    A = fx(newx)/fx(currentx)
    if(runif(1)<A){
      x[i] = newx       # accept move with probabily min(1,A)
    } else {
      x[i] = currentx        # otherwise "reject" move, and stay
where we are
    }
  }
  hist(x,breaks=100,  freq=FALSE,main="Metropolis-Hastings Sampling
",xlab="X",ylim=c(0,0.35))
  plot(fx2,0,8,main="True Distribution",xlab="X",ylab="Density",yl
im=c(0,0.35))

#Example 5-7
  gibbs<-function (n, r)
  {
    mat <- matrix(ncol = 2, nrow = n)
    x <- 0
    y <- 0
    mat[1, ] <- c(x, y)
    for (i in 2:n) {
      x <- rnorm(1, r * y, sqrt(1 - r^2))
      y <- rnorm(1, r * x, sqrt(1 - r^2))
      mat[i, ] <- c(x, y)
    }
    mat
  }
  bvn<-gibbs(10000,0.98)
  par(mfrow=c(3,2))
  plot(bvn,col=1:10000)
```

```
  plot(bvn,type="l")
  plot(ts(bvn[,1]))
  plot(ts(bvn[,2]))
  hist(bvn[,1],40)
  hist(bvn[,2],40)
  par(mfrow=c(1,1))

#Example 5-8
  fx = function(x){
    if(x<0){
      return(0)}
    else {
      return( x*x*exp(-x))
    }
  }
  fx2 =function(x){
    return(fx(x)/2)
  }
  n = 10
  x = rep(0,50000)
  y = rep(0,10)
  z = rep(0,10)
  x[1] = 1      #starting value
  for(i in 2:50000){
    currentx = x[i-1]
    for(j in 1:10){
      y[j]=currentx + rnorm(1,mean=0,sd=1)
    }
    sumy <- 0
    for(k in 1:10){
      sumy <- sumy + fx(y[k])
    }
    ynorm = rep(0,10)
    for(k in 1:10){
      ynorm[k]=fx(y[k])/sumy
    }
    itemy <-  which(rmultinom(1, 1, ynorm) == 1)
    for(j in 1:9){
      z[j] <- y[itemy]+rnorm(1,mean=0,sd=1)
    }
    z[10] <- currentx
    sumz <- 0
    for(k in 1:10){
      sumz <- sumz + fx(z[k])
    }
    alphay <- sumy/sumz
    if(runif(1) < alphay){
      x[i]=y[itemy]
    }
      else{
```

```
        x[i]=currentx
    }
  }
  hist(x,breaks=100,  freq=FALSE,main="Multiple-try Metropolis-
Hastings Sampling",xlab="X",ylim=c(0,0.35))
  plot(fx2,0,8,main="True Distribution",xlab="X",ylab="Density",yl
im=c(0,0.35))

  #Example 5-9
  fx <- function(x){
      return(exp(-2*abs(x-6)))
  }
  n=10000
  x <- array(runif(n,1,11))
  x.f <- apply(x,1,fx)
  10*sum(x.f)/n
  fx2 <- function(x){
    return(exp(-2*abs(x-6))/(1/sqrt(2*pi)*exp(-((x-6)**2)/2)))
  }
  y <- array(rnorm(n,6,1))
  y.f <- apply(y,1,fx2)
  10*sum(y.f)/n
```

References

1 Kroese, D.P., Brereton, T., Taimre, T., and Botev, Z.I. (2014). Why the Monte Carlo method is so important today. *WIREs Computational Statistics* 6: 386–392. doi:10.1002/wics.1314.

2 Lewis, P.A.W., Goodman, A.S., and Miller, J.M. (2010). A pseudo-random number generator for the system/360. *Ibm Systems Journal* 8 (2): 136–146.

3 Keller, A., Heinrich, S., and Niederreiter, H. (eds.) (2007). *Monte Carlo and Quasi-monte Carlo Methods 2006*. Springer Science & Business Media.

4 Flury, B.D. (1990). Acceptance–rejection sampling made easy. *SIAM Review* 32 (3): 474–476.

5 Kroese, D.P., Taimre, T., and Botev, Z.I. (2011). *Handbook of Monte Carlo Methods*. Hoboken, NJ: John Wiley & Sons. 706.

6 Casella, G. and George, E.I. (1992). Explaining the Gibbs sampler. *The American Statistician* 46 (3): 167–174.

7 Heyde, C.C. (2014). Central limit theorem. Wiley StatsRef: Statistics Reference Online.

8 Villen-Altamirano, M. and Villen-Altamirano, J. (1991). Restart: A method for accelerating rare event simulations. *Queuing Performance & Control in Atm Proceedings of International Telegraphic Congress* 13 (4): 299–301.

9 Villén-Altamirano, J. (1998). Restart method for the case where rare events can occur in retrials from any threshold. *AEU - International Journal of Electronics and Communications* 52 (3): 183–189.

10 Zio, E. (2013). The Monte Carlo simulation method for system reliability and risk analysis, Springer Series in Reliability Engineering, doi10.1007/978-1-4471-4588-2

6

Uncertainty Treatment under Imprecise or Incomplete Knowledge

In most engineering system models, the uncertain behavior of a component or system is captured by the probability mass function (pmf) or probability density function (pdf) of its performance. The probability functions are appropriate to describe the randomness in the behavior, i.e. uncertainty of the objective and aleatory type [1], due to the natural variability or stochasticity of the component or system behavior [2]. Another type of uncertainty that must be accounted for in reliability engineering is due to the incomplete or imprecise knowledge about the component or system behavior [3-8], which then is reflected in its modelling and the associated model parameters estimation. This type of uncertainty is often referred to as subjective or epistemic [1,9].

Traditionally, all uncertainties have been described by probabilities. The typical frequentist representation, considering mainly the randomness features, is the most commonly used approach for uncertainty treatment [10]. Subjective probability is used to express the epistemic uncertainty of unknown frequencies, i.e. the chances [11]. However, this approach is reported to be limited to treat various uncertainties. For example, one may assign a failure probability to an offshore platform based on the assumption that its structure can withstand a certain accidental load; while in real-life situations, the structure could fail at a lower load level, and the preassigned probability could not reflect this uncertainty [12]. To meet the practical demands for uncertainty treatment, different approaches have been developed. This chapter presents some of these: interval, fuzzy numbers, possibility, evidence, and random-fuzzy numbers. The first three approaches are focused on epistemic uncertainty, and the latter two are capable of treating aleatory and epistemic uncertainties. The associated basic arithmetic operations for uncertainty propagation are introduced.

6.1 Interval Number and Interval of Confidence

6.1.1 Definition and Basic Arithmetic Operations

Sometimes in practice, with the lack of additional information, the uncertainty in a parameter is described by experts in terms of an interval of possible values within a

Reliability Analysis, Safety Assessment and Optimization: Methods and Applications in Energy Systems and Other Applications, First Edition. Enrico Zio and Yan-Fu Li.

minimum and a maximum. Using intervals is one practical way to deal with such epistemic uncertainty on the value of the parameter. We use the following symbol to denote an interval number:

$$X = [\underline{x}, \overline{x}] = \{x \in \mathbb{R} \mid \underline{x} \le x \le \overline{x}\}$$

where $\underline{x}$ and $\overline{x}$ ($\underline{x} \le \overline{x}$) are the finite lower and upper bounds of X, respectively. In the rest of this section, we will presume that $\underline{x} \le \overline{x}$. In certain cases, we can have $\underline{x} = -\infty$ and/or $\overline{x} = +\infty$. Typically, the brackets $[\cdot]$ indicate a closed interval. In other cases, we may have half-open intervals: $(\underline{x}, \overline{x}]$ as a left-open interval, $[\underline{x}, \overline{x})$ as a right-open interval, and $(\underline{x}, \overline{x})$ as an open interval.

Interval arithmetic is an arithmetic with operations defined on intervals. A form of interval arithmetic first appeared in the early twentieth century [13]. More modern development of interval arithmetic was initiated by R. E. Moore [14] in 1962. The four basic interval arithmetic operations are presented as follows.

Addition	$X + Y = [\underline{x} + \underline{y}, \overline{x} + \overline{y}]$
Subtraction	$X - Y = [\underline{x} - \overline{y}, \overline{x} - \underline{y}]$
Multiplication	$X \times Y = [\min\{\underline{x}\underline{y}, \underline{x}\overline{y}, \overline{x}\underline{y}, \overline{x}\overline{y}\}, \max\{\underline{x}\underline{y}, \underline{x}\overline{y}, \overline{x}\underline{y}, \overline{x}\overline{y}\}]$
Division	$X / Y = X \times 1 / Y$, where $1 / X = [1/\overline{x}, 1/\underline{x}]$ if $\underline{x} > 0$ or $\overline{x} < 0$

The ranges of the operations above are exactly the ranges of the corresponding real operations. For inversion operation, if one interval includes the value of zero, i.e. $\underline{x} < 0 < \overline{x}$, then we typically have a union of two separated intervals, $1/X = [-\infty, 1/\underline{x}] \cup [1/\overline{x}, +\infty]$, as the inversion of X.

6.1.2 Algebraic Properties

The following properties hold for addition and multiplication.

Commutativity	$X + Y = Y + X$, $X \times Y = Y \times X$
Associativity	$X + (Y + Z) = (X + Y) + Z$, $X \times (Y \times Z) = (X \times Y) \times Z$

Example 6.1 Consider the following intervals:

$X = [-2.45, 5.34]$, $Y = [4.56, 9.13]$ and $Z = [-6.43, -1.95]$.

Then, $X + Y = [2.11, 14.47]$,

$$X - Z = [-0.50, 11.77],$$

$$X / Z = [-2.7385, 1.2564],$$

$$X \times Y \times Z = [-22.3685, 48.7542] \times [-6.43, -1.95] = [-313.4895, 143.8295]$$

$$= [-2.45, 5.34] \times [-58.7059, -8.8920] = [-313.4895, 143.8295].$$

As for the distributive law of ordinary arithmetic:

$$x \times (y + z) = x \times y + x \times z,$$

this does not always hold for intervals. For example, let $X = [2, 3]$, $Y = [1, 2]$, and $Z = [-1, 0]$. Then,

$$X \times (Y + Z) = [2, 3] \times [0, 2] = [0, 6],$$

whereas

$$X \times Y + X \times Z = [2, 6] + [-3, 0] = [-1, 6],$$

However, the distributive law is true as long as the intervals Y and Z have the same sign:

$$X \times (Y + Z) = X \times Y + X \times Z \text{ if } Y \times Z > 0,$$

For example, we have

$$[2, 3] \times ([1, 2] + [0, 1]) = [2, 9] = [2, 3] \times [1, 2] + [2, 3] \times [0, 1].$$

6.1.3 Order Relations

Just as real numbers can be ordered, the interval numbers can be ordered, too, e.g. by <, the relation symbol. This relation is transitive for real numbers: If $x < y$ and $y < z$, then $x < z$ for any $x, y, z \in \mathbb{R}$. A similar and preliminary order relation can be defined for intervals:

$$X < Y \text{ means that } \overline{x} < \underline{y},$$

Then, this order relation will have the transitive property,

$$X < Y \text{ and } Y < Z \Rightarrow X < Z,$$

For example, [-1, 0] < [1, 2] and [1, 2] < [3, 4] gives [-1, 0] < [3, 4].

However, this relation cannot be used to compare a large number of intervals which are overlapping, e.g. [1, 3] and [2, 4].

More generalized order relations have been defined to cope with these situations and used in decision-making contexts. The relation $\leq_{LR}$ [15] is one such relation. It is defined as:

$$X \leq_{LR} Y \text{ iff } \underline{x} \leq \underline{y} \text{ and } \overline{x} \leq \overline{y},$$

$$X <_{LR} Y \text{ iff } X \leq_{LR} Y \text{ and } X \neq Y,$$

which represents the decision maker's preference on the interval with the highest minimum and maximum. It is transitive, reflexive, and antisymmetric and, thus, a partial order. Still, many pairs of intervals cannot be compared to it, e.g. $X = [1, 4]$ and $Y = [2, 3]$.

The relation $\leq_{CW}$ is another order relation defined on the basis of the center and width of the interval [15]. For $X = [\underline{x}, \overline{x}]$, we have $x_C = \frac{\underline{x} + \overline{x}}{2}$ and $x_W = \frac{\overline{x} - \underline{x}}{2}$ as the center and width, respectively. The relation is defined as

$$X \leq_{CW} Y \text{ iff } x_C \leq y_C \text{ and } x_W \geq y_W,$$

$$X <_{CW} Y \text{ iff } X \leq_{CW} Y \text{ and } X \neq Y,$$

which represents the decision maker's preference on the interval with the highest expectation and lowest uncertainty. Also, this relation gives only a partial order and many pairs of intervals cannot be compared by it, e.g. $X = [1, 5]$ and $Y = [2, 3]$.

6.1.4 Interval Functions

Let $f(x)$ denote a real-valued function of a single real-valued variable x. By replacing x with an interval X, the resulting function of the interval can be expressed as follows:

$$f(X) = \{f(x) \mid x \in X\}$$

where $f(X)$ is the image of the set X under the mapping $f(\cdot)$. In case of multiple variables, the function of intervals can be written as follows:

$$f(X_1, \ldots, X_n) = \{f(x_1, \ldots, x_n) \mid x \in X_1, \ldots, x \in X_n\},$$

where $X_1, \ldots, X_n$ are n intervals.

Below are two example functions.

6.1.4.1 Quadratic Function

The real-valued function is

$$f(x) = x^2, x \in \mathbb{R}$$

Then, for $X = [\underline{x}, \overline{x}]$, we have

$$f(X) = \{x^2 \mid x \in X\}$$

$$= \begin{cases} [\underline{x}^2, \overline{x}^2], & \underline{x} > 0 \\ [\overline{x}^2, \underline{x}^2], & \overline{x} < 0 \\ [0, \max\{\underline{x}^2, \overline{x}^2\}], & \underline{x} \leq 0 \leq \overline{x} \end{cases}$$

$X^2 \neq X \times X$. For example, $[-2,1]^2 = [0,4]$ whereas $[-2,1] \times [-2,1] = [-2,4]$. This discrepancy is due to *interval dependency*, which assumes the two intervals are independent when we consider multiplying them.

6.1.4.2 Exponential Function

The real-valued function is

$$f(x) = \exp(x), x \in \mathbb{R},$$

Then, for $X = [\underline{x}, \overline{x}]$, we have

$$f(X) = [\exp(\underline{x}), \exp(\overline{x})],$$

The exponential function is one type of *monotonic* function. For all *monotonic increasing* functions, we have $f(X) = [f(\underline{x}), f(\overline{x})]$ whereas, for all *monotonic decreasing* functions, we have $f(X) = [f(\overline{x}), f(\underline{x})]$.

For more information about interval arithmetic, the readers are referred to the book [16].

6.1.5 Interval of Confidence

The intervals mentioned above are regarded as the *intervals of confidence* if the bounds of the intervals are also uncertain. In these cases, to quantify the uncertainty, we may associate a level of presumption to the interval. For example, we can estimate that the lifetime of a component is 11 years, and we can estimate that its lifetime is between 10 and 12 years. We may assign confidence 0 to [10, 12] and confidence 1 to [11, 11]. The two levels of presumption can be represented by a value in the range [0, 1]. Let α denote the presumption level, $\alpha \in [0,1]$. Based on this, we can have the following important property:

$$\forall \alpha_1, \alpha_2 \in [0,1], \quad \alpha_1 > \alpha_2 \Rightarrow [\underline{x}_{\alpha_1}, \overline{x}_{\alpha_1}] \subset [\underline{x}_{\alpha_2}, \overline{x}_{\alpha_2}],$$

which means that if α decreases, the interval of confidence will never decrease. This brings us to the introduction of another important descriptor of uncertainty: the fuzzy number.

6.2 Fuzzy Number

The concept of a fuzzy number can be presented in different ways. If we follow Kaufmann and Gupta [17], it is understood to be an extension of the interval of confidence (as presented in Section 6.1.5). However, the classical way of defining fuzzy numbers is from the viewpoint of fuzzy set theory [18]. Let A^* be an ordinary set that are characterized in terms of the binary function $\mu_{A^*}(x) \in \{0, 1\}$, which indicates whether the given element belongs to set A^*. In a fuzzy set A, any of its elements x is associated with a membership function value (an extension of the characteristic function of the ordinary set above), $\mu_A(x) \in [0, 1]$, which describes the degree to which the element belongs to A. Such a degree of membership takes values in $[0, 1]$.

Two properties are needed to define a fuzzy number on a fuzzy set: convexity and normality. A fuzzy set is convex if and only if each of its ordinary subsets is convex, i.e. a closed interval of real numbers. The normality requires that the highest membership value on A equals 1. A fuzzy number is, then, a convex and normal fuzzy set of real numbers, denoted as $X_\alpha = \{x \mid \mu_X(x) \geq \alpha\}$. Thus, a fuzzy number can be considered a generalization of the interval of confidence introduced in Section 6.1.5. In Figure 6.1, the subplots show a non-normal convex fuzzy set, a normal non-convex fuzzy set and a normal convex fuzzy set, i.e. a fuzzy number, respectively.

The arithmetic operations of fuzzy numbers can be achieved by using the arithmetic operations of the intervals illustrated in the previous sections, applying them to the intervals membership level α. For example, let X and Y be two fuzzy numbers, and let X_α and Y_α be their intervals of confidence at the membership presumption level $\alpha \in [0, 1]$. Then, we can write

$$X_\alpha + Y_\alpha = [\underline{x}_\alpha, \overline{x}_\alpha] + [\underline{y}_\alpha, \overline{y}_\alpha] = [\underline{x}_\alpha + \underline{y}_\alpha, \overline{x}_\alpha + \overline{y}_\alpha].$$

There is another method that can be used for the arithmetic operations of fuzzy numbers, namely the fuzzy extension principle [19]. By this, we have the following expression for the arithmetic operations between two fuzzy numbers,

$$\mu_{X \otimes Y}(z) = \sup_{x \otimes y} \min(\mu_X(x), \mu_Y(y)),$$

where X and Y are fuzzy numbers and $\otimes$ is any arithmetic operation $\{+, -, \times, \div\}$.

Example 6.2 A discrete fuzzy number usually has the following expression $A = \sum_{i=1}^{\infty} \mu_A(a_i) / a_i$. For two discrete fuzzy numbers $X = 0.2/1 + 1/2 + 0.5/3$ and $Y = 0.3/2 + 1/3 + 0.4/4$, their sum can be obtained according to the extension principle as follows:

$$\begin{aligned} X + Y = {} & 0.2/3 + 0.2/4 + 0.2/5 + \\ & 0.3/4 + 1/5 + 0.4/6 + \\ & 0.3/5 + 0.5/6 + 0.4/7. \end{aligned}$$

Maximizing the presumption level at the same value, then we have

$$X + Y = 0.2/3 + 0.3/4 + 1/5 + 0.5/6 + 0.4/7.$$

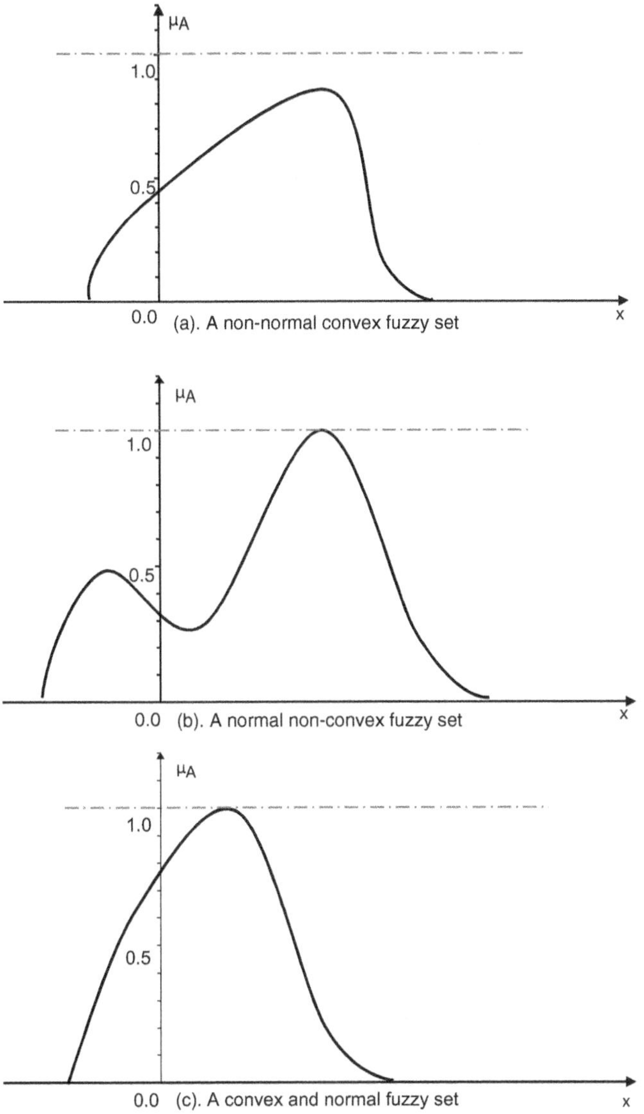

Figure 6.1 (a). A non-normal convex fuzzy set. (b). A normal non-convex fuzzy set. (c). A convex and normal fuzzy set.

6.3 Possibility Theory

The possibility theory is a popular alternative to represent and treat epistemic uncertainty [19]. In this theory, the uncertain number $\tilde{X}$ in the sample space $\Theta \subseteq \Re$ is defined by the possibility distribution function $\pi\colon \Theta \to [0,1]$, such that $\sup_{x\in\Theta} \pi(x) = 1$. For each element $x \in \Theta$, $\pi(x)$ represents the degree of possibility that $\tilde{X}$ takes value x. If there is an element x_i that makes $\pi(x_i) = 0$, then x_i will be regarded as an impossible outcome. On the other hand, if $\pi(x_i)=1$, then x_i will be regarded as a definitely possible outcome, i.e. an unsurprisingly normal, usual outcome [20]. This is a much weaker statement than the situation when probability equals 1, which makes the value x_i certain and the value $x_j \neq x_i$ impossible. It is also known that $\pi_{\tilde{X}}(x)$ is formally equivalent to the fuzzy set $\{\mu(x)/x \mid x \in \Theta\}$ [19]. The two measures of possibility distribution, namely the possibility $\Pi(B)$ and the necessity $N(B)$, are defined as

$$\Pi(B)=\sup_{\{x\in B\}}\pi(x) \text{ and } N(B)=1-\Pi(\bar{B})=\inf_{\{x\in \bar{B}\}}(1-\pi(x)).$$

$\Pi(B)$ indicates to what extent event B is plausible and $N(B)$ indicates to what extent event B is certain. For any pair of events B_1 and B_2, it obeys the following rules:

$$\Pi(B_1\cup B_2)=\max(\Pi(B_1),\, \Pi(B_2)) \text{ and } N(B_1\cap B_2)=\min(N(B_1),N(B_2)).$$

The possibility measures can be linked to probabilities in the following manner [12].

The possibility distribution $\pi(x)$ can also be represented by a nested set of confidence intervals, the α-cuts $[\underline{x}_\alpha,\, \overline{x}_\alpha]=\{x \mid \pi(x)\geq \alpha\}$ of α where $\underline{x}_\alpha$ and $\overline{x}_\alpha$ are respectively the lower and upper limits of the α-cuts, respectively. The degree of certainty of $[\underline{x}_\alpha,\, \overline{x}_\alpha]$ containing the value of $\tilde{X}$ is equal to $N([\underline{x}_\alpha,\, \overline{x}_\alpha])=\min(N(X\geq \underline{x}_\alpha),N(X\leq \overline{x}_\alpha))=1-\alpha$. On the other hand, the α-cuts of a possibility distribution can be interpreted as the probabilistic constraints $P(X\in[\underline{x}_\alpha,\, \overline{x}_\alpha])\geq 1-\alpha$; thus, the possibility distribution is linked to imprecise probability [21]. Then, $N([\underline{x}_\alpha,\, \overline{x}_\alpha])$ corresponds to a lower bound of probability and $\Pi([\underline{x}_\alpha,\, \overline{x}_\alpha])=1$ corresponds to an upper bound of probability.

Example 6.3 Let us consider the opinions given by experts about a certain measurement. They are certain that it varies within the interval [1, 4]. Based on their experience and possibly a few measurements, they suggest the true value of $\tilde{X}$ is most likely to fall into a smaller interval [2, 3]. The possibility distribution, the related possibility measures and the α-cut are depicted in Figure 6.2.

For example, $[\underline{x}_{0.7},\, \overline{x}_{0.7}]=[1.7, 3.3]$ is the set of values for which the possibility distribution function is greater than or equal to 0.7: We conclude that if the event B indicates that the parameters lie in the interval [1.7, 3.3], then $N(B)=0.3\leq P(B)\leq 1=\Pi(B)$.

6.3.1 Possibility Propagation

The possibilistic output $\tilde{Y}$ of a model of possibilistic inputs $\tilde{X}_i$ is often a multivariate function $\tilde{Y}=f(\tilde{X}_1,\tilde{X}_2,\ldots,\tilde{X}_n)$. Given the possibility distributions of the uncertain input variable $\tilde{X}$, it is possible to infer the possibility distribution of $\tilde{Y}$ by means of the α-cut method. For a given input variable $\tilde{X}$, we define the α-cut of $\tilde{X}$ as:

$$X_\alpha=\{x\in U \mid \pi_{\tilde{X}}(x)\geq \alpha, 0\leq \alpha\leq 1\},$$

$$X_\alpha=[\underline{x}_\alpha,\, x_\alpha],$$

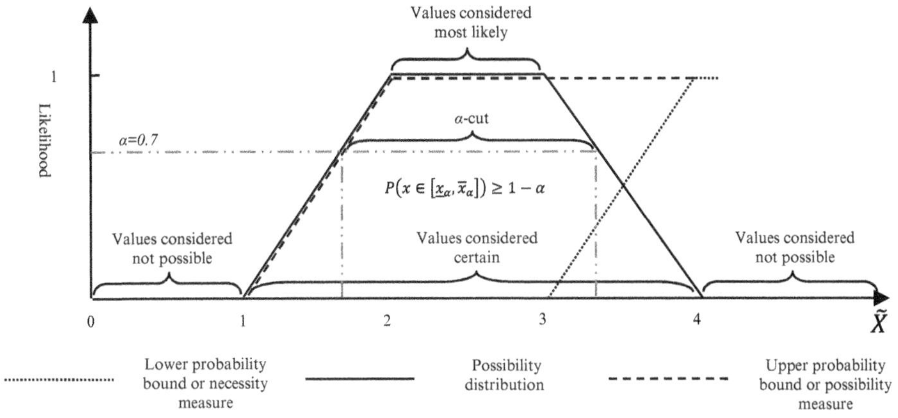

Figure 6.2 Possibility distribution of $\tilde{x}$, and related possibility measures and the α-cut [22].

where U is the universe of discourse of $\tilde{X}$ (i.e. the range of its possible values) and $\underline{x}_\alpha$ and $\overline{x}_\alpha$ are the lower and upper limits of the α-cut, respectively. Given the α-cuts of each uncertain input parameter, the α-cut of the output Y can be obtained as:

$$Y_\alpha = \left[\underline{y}_\alpha, \overline{y}_\alpha\right],$$

$$\underline{Y}_\alpha = \inf f\left(X_{1\alpha}, X_{2\alpha}, \ldots, X_{n\alpha}\right),$$

$$\overline{Y}_\alpha = \sup f\left(X_{1\alpha}, X_{2\alpha}, \ldots, X_{n\alpha}\right),$$

where $X_{i\alpha}$ represents the α-cut of the ith possibilistic input variable. For each α-cut of the output $\tilde{Y}$, the maximum and minimum outputs (upper bound $\overline{Y}_\alpha$, and lower bound $\underline{Y}_\alpha$) are obtained.

6.4 Evidence Theory

The evidence theory, also called Dempster-Shafer theory [23], provides a single framework to treat variability and imprecision separately. Let $\Omega = \{\theta_1, \ldots, \theta_n\}$ denote a finite discrete set of mutually exclusive events, called the frame of discernment. It is assumed that one's total belief induced by a body of evidence can be divided into various portions, each one assigned to a subset A of Ω. To express this, a basic belief assignment (BBA) function is defined on the power set 2^Ω mapping the belief masses onto the events or sets of events:

$$m(\phi) = 0 \text{ and } \sum_{A \subseteq \Omega} m(A) = 1,$$

where ϕ represents the empty set and $m(A)$ is the belief mass that one is willing to commit exactly to A and not to its subsets. For example, let $\Omega = \{1,2\}$: then, $2^\Omega = \{\phi, 1, 2, \Omega\}$, and a BBA function can be defined, for example, $m(1)=0.1$, $m(2)=0.2$, $m(\{1,2\})=0.7$. The subset $A \subseteq \Omega$ is called a focal element if $m(A)>0$ and $\mathcal{F}$ are the set of all focal elements induced by m. The duplet $\mathcal{B}=\langle \mathcal{F},m\rangle$ is referred to as the *body of evidence*.

A Bayesian BBA is a special case where all the focal elements are singletons and the belief masses equal probabilities.

The differences between probability distribution functions and BBAs are twofold: 1) the probability distribution functions are defined on Ω whereas the BBAs are defined on the power set 2^{Ω} and 2) the sub-additivity hypothesis is not required in the evidence theory as it is, instead, in the probability theory.

The evidence theory provides two indicators to quantitatively describe uncertainty with respect to a set A: the belief $Bel(A)$ and the plausibility $Pl(A)$ functions, which are also referred to as belief functions. A portion of belief mass committed to an element A must be committed to any of its subsets: to obtain the total belief in A, one must sum up the belief masses of every subset B of A. The function that accounts for the total belief of A is called belief function:

$$Bel(A)=\sum_{B\subseteq A, \forall B\subseteq 2^{\Omega}} m(B),$$

It is easily verified that the belief in some hypothesis A and the belief in its complement $\bar{A}$ do not necessarily sum to 1. Therefore, $Bel(A)$ does not reveal to what extent one believes in $\bar{A}$ or, dually, to what extent one doubts A. Instead, it is the quantity $Pl(A)$, namely the plausibility of A, which is introduced to define to what extent one fails to doubt in A:

$$Pl(A)=\sum_{B\cap A\neq\phi,\ \forall B\subseteq 2^{\Omega}} m(B),$$

Belief and plausibility have the following relations:

$$Bel(A)=1-Pl(\bar{A}),$$

$$Pl(A)=1-Bel(\bar{A}).$$

$Bel(A)$ gathers the imprecise evidence that asserts A, and $Pl(A)$ gathers the imprecise evidence that does not conflict with A. Therefore, the interval $[Bel(A),Pl(A)]$ contains all probability values induced by the mass distribution $m(A)$ on subset A. The mass distribution m is the generalization of the probability distribution p and the possibility distribution π of uncertain discrete variables (the continuous variables have to be discretized) [24]. The evidence theory, thus, encompasses the probability theory and possibility theory in two ways: 1) when the focal elements are nested, Bel is a necessity measure, that is $Bel=\Pi$ and Pl is a possibility measure, that is $Pl=N$ and 2) when the focal elements are some disjoint intervals, and Bel and Pl are both probability measures, that is $Bel=Pl=P$.

From the above section, we can calculate the belief and plausibility functions from the BBA. On the other hand, if we know belief or plausibility, then we will be able to calculate the BBA using the following formula:

$$m(A) = \sum_{B \subseteq A} (-1)^{|A-B|} Bel(B),$$

where $|A-B|$ is the cardinality of the difference of set A from set B.

Example 6.4 Let $A = \{a_1, a_2, a_3\}$ and $m(\{a_1, a_2\}) = 0.3, m(\{a_1\}) = 0.1,\ m(\{a_2, a_3\}) = 0.2, m(A) = 0.4$. The focal set of this BBA is $\mathcal{F} = \{\{a_1\}, \{a_1, a_2\}, \{a_2, a_3\}, A\}$.

We can compute the belief and plausibility of any subset of A. For example, the belief in $\{a_1, a_2\}$ is

$$Bel(\{a_1, a_2\}) = m(\{a_1, a_2\}) + m(\{a_1\}) = 0.4.$$

The plausibility of $\{a_1, a_2\}$ is

$$Pl(\{a_1, a_2\}) = m(\{a_1, a_2\}) + m(\{a_1\}) + m(A) = 0.8.$$

Given the belief values, to compute the BBA of $\{a_1, a_2\}$, we have the following formula:

$$m(\{a_1, a_2\}) = (-1)^1 \times 0.1 + (-1)^0 \times 0.4 = 0.3.$$

In the end, we can build the following table for the BBAs, belief and plausibility of all subsets of A.

Table 7.1 The BBAs, belief and plausibility of subsets of A.

Set	*m*	*Bel*	*Pl*
ϕ	0	0	0
$\{a_1\}$	0.1	0.1	0.8
$\{a_2\}$	0	0	0.9
$\{a_3\}$	0	0	0.6
$\{a_1, a_2\}$	0.3	0.4	0.8
$\{a_1, a_3\}$	0	0.1	0.7
$\{a_2, a_3\}$	0.2	0.2	0.9
A	0.4	1	1

6.4.1 Data Fusion

The data fusion method was developed to combine the evidence from different sources of uncertainty. Because a piece of evidence can be possibilistic (fuzzy) or probabilistic, evidence theory provides a framework to fuse the different uncertainties. Let $\mathcal{B}_1 = \mathcal{F}_1, m_1$ and $\mathcal{B}_2 = \mathcal{F}_2, m_2$ denote two bodies of evidence where $\mathcal{F}_1$ and $\mathcal{F}_2$ are the focal sets of the same universe U, induced by m_1 and m_2, respectively. The conflict occurs whenever the focal elements have no overlap, i.e. $A_1 \cap A_2 = \varnothing$. Thus, the total conflict of the two evidence bodies is defined as follows:

$$\kappa = \sum_{A_1 \cap A_2 = \varnothing} m_1(A_1) m_2(A_2)$$

where $A_1 \in \mathcal{F}_1$ and $A_2 \in \mathcal{F}_2$. Then, we can create a fused body of evidence $\mathcal{B}_f = \mathcal{F}_f, m_f$ from the two evidence bodies $\mathcal{B}_1$ and $\mathcal{B}_2$. First, the focal set is defined as follows:

$$\mathcal{F}_f = \{A_1 \cap A_2 \mid A_1 \cap A_2 \neq \varnothing,\ A_1 \in \mathcal{F}_1 \text{ and } A_2 \in \mathcal{F}_2\}.$$

Then, the BBA m_f is defined as

$$m_f(B) = \frac{\sum_{A_1 \cap A_2 = C} m_1(A_1) m_2(A_2)}{1 - \kappa}$$

The normalization factor $1 - \kappa$ is introduced at the denominator to ensure m_f adds up to 1. For more details about evidence theory, the readers can refer to the book [25].

6.5 Random-fuzzy Numbers (RFNs)

Random-fuzzy numbers (RFNs) were first introduced by Kaufmann and Gupta [17] as a tool to jointly express epistemic and aleatory uncertainties. Later, RFN Cooper, et al. [26] and Baudrit, et al. [24] extended it to hybrid uncertainty propagation in the area of risk analysis. Given the monotonicity of the cumulative distribution functions (cdfs) of random variables and the nestedness of the possibility distribution functions of the fuzzy numbers, the formal definition of RFN proposed by Ferson and Ginzburg [27] is presented as follows.

Definition (Ferson and Ginzburg [27]) Let F denote the set of all cdfs defined on the real number set $\mathbb{R}$, and each element $F \in \boldsymbol{F}$ is an onto function $F : \mathbb{R} \to [0,1]$ such that $F(x_1) \geq F(x_2)$ whenever $x_1 > x_2$. An RFN is a set of closed intervals, each characterized by a pair of functions from $\boldsymbol{F}$:

$$H : [0,1] \to \boldsymbol{F} \times \boldsymbol{F} : \alpha \mapsto [\underline{F}_\alpha, \overline{F}_\alpha]$$

such as for $\alpha_1, \alpha_2 \in [0,1]$, $\overline{F}_{\alpha_1}(x) \geq \overline{F}_{\alpha_2}(x) \geq \underline{F}_{\alpha_2}(x) \geq \underline{F}_{\alpha_1}(x)$ wherenever $\alpha_1 < \alpha_2$, where α_1 and α_2 represent fuzzy membership values of x.

Example 6.5 Figure 6.3 (a) depicts the three-dimensional representation of an RFN. The x-axis is the real number line, the F-axis gives the cumulative probability values, and the π-axis contains the possibility values. The shaded area at the $\alpha \in (0,1)$ level includes all the closed probability intervals, limited by $\underline{F}_\alpha$ as the lower bound and $\overline{F}_\alpha$ as the upper bound. Figure 6.3(b) depicts the intersection of the RFN with the plane $F(x) = p$, which is essentially a fuzzy number. Figure 6.3(c) shows a two-dimensional representation of the RFN from Figure 6.1(a), and its α level probability intervals. Figure 6.3(d) depicts the intersection of the RFN with the plane $F(x) = p$, from Figure 6.3(c), in a two-dimensional representation.

6.5.1 Universal Generating Function (UGF) Representation of Random-fuzzy Numbers

We first recall the UGF for a discrete random variable X, as

$$u_X(z) = \sum_{j=0}^{J} p_j z^{x_j} \tag{6.1}$$

where z is the base of z-transform, J is the total number of realizations of X, x_j is the j-th realization of X, and p_j is the probability mass attached to x_j and satisfying $\sum_{j=0}^{J} p_j = 1$.

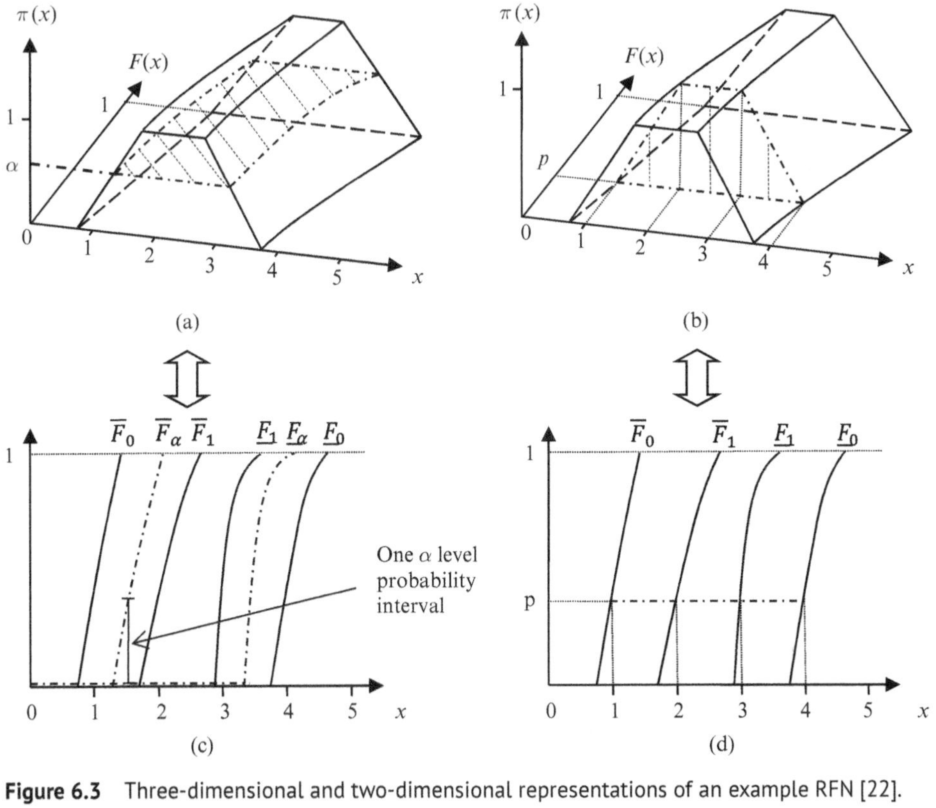

Figure 6.3 Three-dimensional and two-dimensional representations of an example RFN [22].

The u-function is useful in representing the probability distribution function of discrete random variables because it preserves some basic properties of the moment-generating function, which uniquely determines its probability distribution function [28].

Definition The u-function of a discrete RFN $\ddot{X}$, denoted by $u_{\ddot{X}}(z)$, can be written as follows:

$$u_{\ddot{X}}(z) = \sum_{j=0}^{J} p_j z^{\tilde{X}_j} = \sum_{j=0}^{J} p_j z^{[\underline{x}_j, \overline{x}_j]_\alpha} \tag{6.2}$$

This definition satisfies the basic property of the UGF: The coefficient and exponent are not necessarily scalar variables but can be other mathematical objects (e.g. vector, matrix) [28]. It is seen that Equation (6.1) is the special case of Equation (6.2): if all the exponents of z in Equation (6.2) are crisp values (i.e. sufficient information is collected to eliminate the imprecision in state values), then Equation (6.2) will reduce to Equation (6.1). On the other hand, if there is only one coefficient of z, equal to 1, then Equation (6.2) will reduce to the following expression:

$$u_{\tilde{X}}(z) = z^{\tilde{X}} = z^{[\underline{x}_\alpha, \overline{x}_\alpha]} \tag{6.3}$$

which is the u-function of a pure fuzzy number. Recall that $\pi_{\tilde{X}}(x)$ can be uniquely determined by its α-cut set $[\underline{x}_\alpha, \overline{x}_\alpha]$; therefore, Equation (6.3) defines a one-to-one correspondence to $\tilde{X}$. For example, the u-function of the fuzzy number depicted in Figure 6.2 is $z^{[1+\alpha, 4-\alpha]}$.

6.5.2 Hybrid UGF (HUGF) Composition Operator

Because RFN treats the two types of uncertainties separately, the composition operator of hybrid UGF has the properties of probabilistic UGF composition operator [29] and of the fuzzy extension principle [19]. In the following three cases, we will show that the conventional UGF composition operator $\otimes_f$ is applicable to hybrid UGF compositions if its structure function $f(\cdot)$ supports fuzzy arithmetic operations.

Case 1: *$\otimes_f$ between the u-functions of two fuzzy variables $\tilde{X}_1$ and $\tilde{X}_2$,*

$$u_{\tilde{X}_1}(z) \otimes_f u_{\tilde{X}_2}(z) = z^{f(\tilde{X}_1, \tilde{X}_2)}$$

The extension principle [19] reads that $\pi_{\tilde{Y}}(y) = \sup_{y=f(x_1,x_2)} \min\left(\pi_{\tilde{X}_1}(x_1), \pi_{\tilde{X}_2}(x_2)\right)$. For example, if we have $\tilde{X}_1 = [1+\alpha, 4-\alpha]$ and $\tilde{X}_2 = [2+\alpha, 3-\alpha]$, then the u-function of the denominator will be able to be written as:

$$u_{\tilde{X}_1}(z) \otimes_\times u_{\tilde{X}_2}(z) = z^{\tilde{X}_1 \times \tilde{X}_2} = z^{[(1+\alpha)\times(2+\alpha), (4-\alpha)\times(3-\alpha)]}.$$

The fuzzy arithmetic presumes the total dependence between the α-cuts [24].

Case 2: $\otimes_f$ *between one random variable X_1 and one fuzzy variable $\tilde{X}_2$,*

$$u_{X_1}(z)\otimes_f u_{\tilde{X}_2}(z)=\sum_{j_1=0}^{J_1} p_{1j_1} z^{f\left(X_{1j_1},\tilde{X}_2\right)}$$

For example, suppose that X_3 has three states (0, 0.2, 0.8) with the probability vector (0.4, 0.4, 0.2) and $\tilde{X}_1=\left[1+\alpha,4-\alpha\right]$; then, the outcome of this term can be written as:

$$u_{X_3}(z)\otimes_\times u_{\tilde{X}_1}(z)=0.4z^0+0.4z^{\left[0.2(1+\alpha),\,0.2(4-\alpha)\right]}+0.2z^{\left[0.8(1+\alpha),\,0.8(4-\alpha)\right]}.$$

Case 3: $\otimes_f$ *between two random fuzzy variables $\ddot{X}_1$ and $\ddot{X}_2$,*

$$u_{\ddot{X}_1}(z)\otimes_f u_{\ddot{X}_2}(z)=\sum_{j_1=0}^{J_1}\sum_{j_2=0}^{J_2} p_{1j_1}p_{2j_2} z^{f\left(\tilde{X}_{1j_1},\,\tilde{X}_{2j_2}\right)}$$

For example, we have the following operation for the addition of two RFNs:

$$\begin{aligned}
u_{\ddot{X}_1}(z)\otimes_+ u_{\ddot{X}_2}(z)&=\left(0.4z^0+0.4z^{\left[0.2(1+\alpha),0.2(4-\alpha)\right]}+0.2z^{\left[0.8(1+\alpha),0.8(4-\alpha)\right]}\right)\otimes_+\\
&\left(0.4z^0+0.4z^{\left[0.2(1+\alpha)(2+\alpha),0.2(4-\alpha)(3-\alpha)\right]}+0.2z^{\left[0.8(1+\alpha)(2+\alpha),0.8(4-\alpha)(3-\alpha)\right]}\right)\\
&=0.16z^0+0.16z^{\left[0.2(1+\alpha)(2+\alpha),0.2(4-\alpha)(3-\alpha)\right]}+0.08z^{\left[0.8(1+\alpha)(2+\alpha),0.8(4-\alpha)(3-\alpha)\right]}+\\
&0.16z^{\left[0.2(1+\alpha),0.2(4-\alpha)\right]}+0.16z^{\left[0.2(1+\alpha)(3+\alpha),0.2(4-\alpha)(4-\alpha)\right]}+\\
&0.08z^{\left[0.2(1+\alpha)(9+4\alpha),0.2(4-\alpha)(13-4\alpha)\right]}+0.08z^{\left[0.8(1+\alpha),0.8(4-\alpha)\right]}+\\
&0.08z^{\left[0.2(1+\alpha)(6+\alpha),0.2(4-\alpha)(7-\alpha)\right]}+0.04z^{\left[0.8(1+\alpha)(3+\alpha),0.8(4-\alpha)(4-\alpha)\right]}.
\end{aligned}$$

In general, the HUGF composition operator of n u-functions, i.e. components, is defined as follows:

$$\otimes_f\left(u_{\ddot{X}_1}(z),u_{\ddot{X}_2}(z),\ldots,u_{\ddot{X}_n}(z)\right)=\sum_{j_1=0}^{J_1}\cdots\sum_{j_n=0}^{J_n}\prod_{i=1}^{n}p_{ij_i}z^{f\left(\tilde{X}_{1j_1},\tilde{X}_{2j_1},\ldots,\tilde{X}_{nj_n}\right)}$$

For the case of two arguments, the following two interchangeable notations can be used:

$$\otimes_f\left(u_{\ddot{X}_1}(z),u_{\ddot{X}_2}(z)\right)=u_{\ddot{X}_1}(z)\otimes_f u_{\ddot{X}_2}(z).$$

Two basic properties of $\otimes_f$, namely the associative and commutative properties, are used for reducing the computation time of uncertainty propagation.

If the function $f(\cdot)$ possesses the associative property for any of its component, then $\otimes_f$ will possess this property:

$$\otimes_f\left(u_{\ddot{X}_1}(z),\ldots,u_{\ddot{X}_i}(z),u_{\ddot{X}_{i+1}}(z),\ldots,u_{\ddot{X}_n}(z)\right)=$$
$$\otimes_f\left(\otimes_f\left(u_{\ddot{X}_1}(z),\ldots,u_{\ddot{X}_i}(z)\right),\otimes_f\left(u_{\ddot{X}_{i+1}}(z),\ldots,u_{\ddot{X}_n}(z)\right)\right).$$

If the function $f(\cdot)$ possesses the commutative property for any of its component, then $\otimes_f$ will possess this property:

$$\otimes_f\left(u_{\ddot{X}_1}(z),\ldots,u_{\ddot{X}_i}(z),u_{\ddot{X}_{i+1}}(z),\ldots,u_{\ddot{X}_n}(z)\right)=$$
$$\otimes_f\left(u_{\ddot{X}_1}(z),\ldots,u_{\ddot{X}_{i+1}}(z),u_{\ddot{X}_i}(z),\ldots,u_{\ddot{X}_n}(z)\right).$$

These properties are useful in reducing the computation time. By applying these two properties, the elementary random and fuzzy variables might be separated:

$$\otimes_f\left(u_{X_1}(z),\ldots,u_{\tilde{X}_l}(z),u_{\tilde{X}_1}(z),\ldots,u_{X_m}(z)\right)=$$
$$\otimes_f\left(\otimes_f\left(u_{X_1}(z),\ldots,u_{X_m}(z)\right),\otimes_f\left(u_{\tilde{X}_1}(z),\ldots,u_{\tilde{X}_l}(z)\right)\right).$$

In this way, the fuzzy numbers can be processed prior to combination with the probabilistic variables, which involves multiplying the polynomials. Thanks to the total dependence between the α-cuts, the convolution type of computation can be avoided.

For further details on uncertainty treatment methods and their application to system RAMS, the interested reader can consult the book in reference [30].

6.6 Exercises

1) Prove the distributive law for interval numbers.
2) For two interval numbers, prove that if $X \leq_{CW} Y$ and $X \leq_{LR} Y$ hold, then $X = Y$.
3) For two fuzzy numbers X and Y and given the respective membership functions

$$\mu_X(x)=\begin{cases} 0, & x\leq -5,\\ \dfrac{x}{3}+\dfrac{5}{3}, & -5\leq x\leq -2,\\ -\dfrac{x}{3}+\dfrac{1}{3}, & -2\leq x\leq 1,\\ 0, & 1\leq x, \end{cases}$$

and

$$\mu_Y(y)=\begin{cases}0, & y\leq -3,\\ \dfrac{y}{7}+\dfrac{3}{7}, & -3\leq y\leq 4,\\ -\dfrac{y}{8}+\dfrac{12}{8}, & 4\leq y\leq 12,\\ 0, & 12\leq y,\end{cases}$$

compute the membership functions of $X-Y$ and $X \div Y$.

4) For two bodies of evidence, $\mathcal{B}_1=\mathcal{F}_1,m_1$ and $\mathcal{B}_2=\mathcal{F}_2,m_2$, we have the following focal sets and BBA functions:

$\mathcal{F}_1=\{\{a_3\},\{a_1,a_2\},\{a_2,a_3\},\{a_1,a_2,a_3\}\}$ and $m_1(\mathcal{F}_1)=\{0.1,0.3,0.2,0.4\}$;

$\mathcal{F}_2=\{\{a_1\},\{a_1,a_2\},\{a_1,a_2,a_3\}\}$ and $m_2(\mathcal{F}_2)=\{0.3,0.2,0.5\}$.

Compute the fused body of evidence for all subsets of $A=\{a_1,a_2,a_3\}$.

5) For the following fuzzy random variable

$$u_{\tilde{X}}(z)=\sum_{j=0}^{J}p_j z^{\tilde{X}_j}=\sum_{j=0}^{J}p_j z^{[\underline{x}_j,\bar{x}_j]_\alpha}$$

what is its mean and variance?

References

1 Apostolakis, G.E. (1990). The concept of probability in safety assessments of technological systems. *Science* 250 (4986): 1359–1364.

2 Montgomery, D.C. and Runger, G.C. (2010). *Applied Statistics and Probability for Engineers*, 5th e. Hoboken, NJ: John Wiley & Sons.

3 Li, Y.F. and Zio, E. (2012). Uncertainty analysis of the adequacy assessment model of a distributed generation system. *Renewable Energy* 41: 235–244.

4 Singer, D. (1990). A fuzzy set approach to fault tree and reliability analysis. *Fuzzy Sets and Systems* 34: 145–155.

5 Lin, C.H., Ke, J.C., and Huang, H.I. (2012). Reliability-based measures for a system with an uncertain parameter environment. *International Journal of Systems Science* 43 (6): 1146–1156.

6 Wang, A.S., Luo, Y., Tu, G.Y., and Liu, P. (2011). Quantitative evaluation of human-reliability based on Fuzzy-clonal selection. *IEEE Transactions on Reliability* 60 (3): 517–527.

7 Cai, K.Y. (1996). *Introduction to Fuzzy Reliability*. Kluwer Academic Pub-lishers.

8 Chen, S.M. (1994). Fuzzy system reliability analysis using fuzzy number arithmetic operations. *Fuzzy Sets and Systems* 64: 31–38.

9 Helton, J.C. (2004). Alternative representations of epistemic uncertainty. *Reliability Engineering & System Safety* 85 (1-3): 1–10.

10 Paté-Cornell, M.E. (1996). Uncertainties in risk analysis: Six levels of treatment. *Reliability Engineering & System Safety* 54 (2-3): 95–111.

11 Kaplan, S. and Garrick, B.J. (1981). On the quantitative definition of risk. *Risk Analysis* 1 (1): 11–27.

12 Aven, T. and Zio, E. (2011). Some considerations on the treatment of uncertainties in risk assessment for practical decision making. *Reliability Engineering & System Safety* 96 (1): 64–74.

13 Burkill, J.C. (1924). Functions of intervals. *Proceedings of the London Mathematical Society* 2 (1): 275–310.

14 Moore, R.E. (1962). Interval arithmetic and automatic error analysis in digital computing. DTIC Document.

15 Ishibuchi, H. and Tanaka, H. (1990). Multiobjective programming in optimization of the interval objective function. *European Journal of Operational Research* 48 (2): 219–225.

16 Moore, R.E., Kearfott, R.B., and Cloud, M.J. (2009). *Introduction to Interval Analysis*. Philadelphia, PA: SIAM.

17 Kaufmann, A. and Gupta, M.M. (1985). *Introduction to Fuzzy Arithmetic: Theory and Applications*. New York City: Van Nostrand Reinhold.

18 Zadeh, L.A. (1996). Fuzzy sets. In: *Fuzzy Sets, Fuzzy Logic, and Fuzzy Systems: Selected Papers by Lotfi A Zadeh*, 394–432. World Scientific, River Edge, NJ.

19 Dubois, D., Nguyen, H.T., and Prade, H. (2000). Possibility theory, probability and fuzzy sets: Misunderstandings, bridges and gaps. In: *Fundamentals of Fuzzy Sets* (ed. D. Dubois and H. Prade), 343–438. Boston, MA: Kluwer.

20 Dubois, D. (2006). Possibility theory and statistical reasoning. *Computational Statistics & Data Analysis* 51: 47–69.

21 de Cooman, G. and Aeyels, D. (1999). Supremum-preserving upper probabilities. *Information Sciences* 118: 173–212.

22 Li, Y.F., Ding, Y., and Zio, E. (Mar 2014). Random Fuzzy extension of the universal generating function approach for the reliability assessment of multi-state systems under aleatory and epistemic uncertainties. *IEEE Transactions on Reliability* 63 (1): 13–25. doi:10.1109/tr.2014.2299031.

23 Shafer, G. (1976). *A Mathematical Theory of Evidence*. Princeton, NJ: Princeton Univ. Press.

24 Baudrit, C., Dubois, D., and Guyonnet, D. (2006). Joint propagation of probabilistic and possibilistic information in risk assessment. *IEEE Transactions on Fuzzy Systems* 14 (5): 593–608.

25 Yager, R.R. and Liu, L. (2008). *Classic Works of the Dempster-Shafer Theory of Belief Functions*, 1 e. (Studies in Fuzziness and Soft Computing), 806. Berlin Heidelberg: Springer-Verlag.

26 Cooper, J.A., Ferson, S., and Ginzburg, L. (1996). Hybrid processing of stochastic and subjective uncertainty data. *Risk Analysis* 16 (6): 785–791.

27 Ferson, S. and Ginzburg, L.R. (1995). Hybrid arithmetic. In: *ISUMA-NAFIPS*, 619–623. Los Alamitos, CA.

28 Lisnianski, A. and Levitin, G. (2003). *Multi-state System Reliability: Assessment, Optimization and Applications*. Singapore: World Scientific.

29 Ushakov, I. (1986). Universal generating function. *Soviet Journal of Computer Systems Science* 24 (5): 118–129.

7

Applications

This chapter contains two case studies that make use of the uncertainty quantification and computation tools introduced in all the previous chapters of Part II. The first case study is about the reliability assessment of a distributed power generation system under hybrid uncertainties. The fuzzy number, possibility distribution, and evidence function are implemented; the uncertainty propagation algorithm is introduced. The detailed version of this case study can be found in [1]. The second case study is about the degradation modelling of a nuclear component subject to multiple failure modes. The multi-state system (MSS) models, Markov processes, and Monte Carlo simulation (MCS) algorithm are presented for this application example. The details about this case study can be found in [2].

7.1 Distributed Power Generation System Reliability Assessment

7.1.1 Reliability of Power Distributed Generation (DG) System

We present a model for the reliability assessment of a representative distributed power generation system, which consists of a number of power generation and consumption units. The generation units include renewable generators, e.g. solar generators, wind turbines, electric vehicles (EVs) and the conventional power source by way of transformers (Figure 7.1). The transmission lines are often left out of consideration in the reliability assessment studies [3,4]. The consumption units can be different types of loads, e.g. residential, commercial, and industrial loads [5].

In the power engineering domain, reliability is defined somewhat differently. In many cases, reliability assessment is conducted in the form of power adequacy (P_A) assessment, which focuses on evaluating the sufficiency of generation facilities within the system to satisfy the consumer demand [6] (i.e. power generation P_G exceeding load power consumption P_L):

$$P_A = P_G - P_L. \tag{7.1}$$

Reliability Analysis, Safety Assessment and Optimization: Methods and Applications in Energy Systems and Other Applications, First Edition. Enrico Zio and Yan-Fu Li.

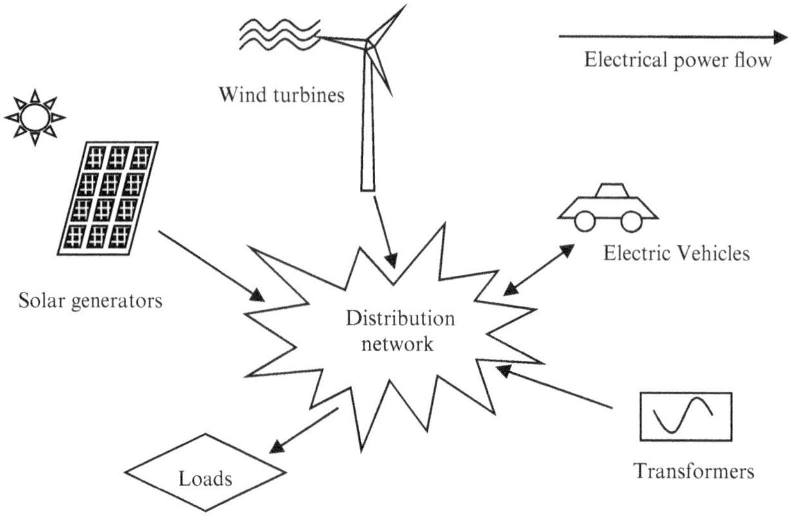

Figure 7.1 Conceptual diagram of the representative distributed generation system [1].

The power adequacy quantity is related to the reliability of the MSS introduced in Chapter 3, where system reliability is described considering the adequacy index P_A as $\Pr\left(P_A \geq 0\right)$.

Power generation P_G consists of two parts: power from the transmission system, P_T, and power from the distributed generators, P_{DG}:

$$P_G = P_T + P_{DG}. \tag{7.2}$$

Considering the distributed generators units of the representative distributed power generation system in Figure 7.1, the compound power output P_{DG} is

$$P_{DG} = P_S + P_W + P_{EV}, \tag{7.3}$$

where $P_S = \sum_{i=1}^{m_S} P_i^S$, $P_W = \sum_{i=1}^{m_W} P_i^W$, and $P_{EV} = \sum_{i=1}^{m_{EV}} P_i^{EV}$ are the power outputs from the set of m_S solar generators, m_W wind turbines, and m_{EV} EVs, respectively, with P_i^S, P_i^W, and P_i^{EV} being the power outputs from the individual units. The value of P_{EV} is negative when the EV group is charging batteries (i.e. consuming power from the network).

7.1.2 Energy Source Models and Uncertainties

The power function of the ith solar generator can be written as

$$P_i^S = g_S\left(s_i, \theta_i^S\right). \tag{7.3}$$

Solar irradiation s_i is typically modeled by a probabilistic distribution (e.g. beta distribution) because the historical solar irradiation data is often sufficient and accessible to justify such representation [7]. The operation parameters θ_i^S (detailed definitions can be found in [1]) are normally provided as deterministic values. However, due to the changing operation conditions, they are not fixed values but little data are available to build probability distributions for them. Consequently, experts' judgments are used to estimate the operation parameters, with inevitable imprecision. To capture this, possibility distributions can be used.

The wind turbines model have a similar description of the uncertainties as the solar generators model. The power function of the ith wind turbine is written as

$$P_i^W = g_W\left(v_i, \theta_i^W\right). \tag{7.4}$$

Wind speed v_i is modeled by a probabilistic distribution (e.g. Weibull distribution) because the historical wind speed data are sufficient and accessible to support such representation of uncertainty. The operation parameters θ_i^W (details can be found in [1]) of the ith wind turbine model are considered 'coefficients'. Similar to the solar generation parameters, we adopt probability distributions for the wind turbine operation parameters.

All EVs distributed on the network are treated as a single aggregation with three possible power output states: charging ($P_{EV} < 0$), disconnection ($P_{EV} = 0$), and discharging ($P_{EV} > 0$). Different from solar and wind generators, EV power outputs are primarily influenced by the activities of the drivers, who can decide the amount of energy to be exchanged with the grid and the timing/location for the exchange. Due to privacy issues, gathering informative operation data for each EV might be difficult, so the estimation of the model parameters relies on expert judgment and knowledge of the drivers' behavior, which is imprecise. Then, the possibilistic distribution is chosen to model the uncertainties in EV power.

As for transformers, the operation has two types of uncertainties: fluctuations of the grid and hardware degradation. Due to the inherent fluctuations in the grid, the power output of the transformer in its working state varies from 80% to 100% of its capacity. As for the degradation and failure mechanisms of the transformers, they have been extensively studied and sufficient data exist to estimate the parameters of probabilistic distributions for describing them. Finally, the real-time load values are monitored by the metering devices installed at the load points, and data are available to establish a probabilistic representation of the associated uncertainties.

Table 7.1 summarizes the aforementioned uncertainties. The overall adequacy assessment model of the distributed generators system can be written as

$$P_A = f\left(s_1, \ldots, s_{m_S}, v_1, \ldots, v_{m_W}, P_T, P_L, \tilde{P}_{EV}, \tilde{\theta}_1^S, \ldots, \tilde{\theta}_{m_S}^S, \tilde{\theta}_1^W, \ldots, \tilde{\theta}_{m_W}^W\right) \tag{7.5}$$

where the possibilistic variables are denoted by the symbol (~). The system adequacy output is a function of aleatory and epistemic uncertain variables and parameters.

Table 7.1 Different uncertainties of the energy models of the distributed generators system [1].

Component	Parameter	Source of uncertainty	Type of Information available	Uncertainty representation
Solar generator	Solar irradiation	Irradiation variability	Historical data	Probabilistic (e.g. Beta)
	Operation parameters	Incomplete knowledge	Experts' judgments, users' experiences	Possibilistic
Wind turbine	Wind speed	Speed variability	Historical data	Probabilistic (e.g. Weibull)
	Operation parameters	Incomplete knowledge	Experts' judgments, users' experiences	Possibilistic
EV aggregation	Power output	Incomplete knowledge, subjective decisions	Experts' judgments, users' experiences	Possibilistic
Transformer	Grid power	Power fluctuations	Historical data	Probabilistic
	Time to failure	Mechanical degradation/failure data	Historical data	Probabilistic
Load	Load value	Consumption variability	Historical data	Probabilistic

7.1.3 Algorithm for the Joint Propagation of Probabilistic and Possibilistic Uncertainties

Consider a general power adequacy model $Y = f\left(X_1,\ldots X_k, \tilde{X}_{k+1},\ldots,\tilde{X}_n\right)$ of n uncertain variables $X_i, i = 1,\ldots,n$, ordered in such a way that the first k variables are described by probability distributions $\left(p_{X_1}(x),\ldots,p_{X_k}(x)\right)$, and the last n-k variables are possibilistic and represented by possibility distributions $\left(\pi_{\tilde{X}_{k+1}}(x),\ldots,\pi_{\tilde{X}_n}(x)\right)$. The propagation of the hybrid uncertainty can be performed by MCS combined with the extension principle of fuzzy set theory by means of the following two major steps [8]:

1) Repeat MCS to process the uncertainty in the probabilistic variables.
2) Analyze fuzzy intervals for treating the uncertainty in the possibilistic variables.

The detailed algorithm [9] to calculate the fuzzy random output can be summarized as follows:

For i = 1, 2, ..., m (the outer loop processing aleatory uncertainty), do the following:

1) Sample the ith realization $\left(x_1^i, x_2^i,\ldots,x_k^i\right)$ of the probabilistic variable vector $\left(X_1, X_2,\ldots,X_k\right)$.
2) For $\alpha = 0, \Delta\alpha, 2\cdot\Delta\alpha,\ldots,1$ (the inner loop processing epistemic uncertainty), $\Delta\alpha$ is the step size, e.g. $\Delta\alpha$=0.05), do:

2.1 Calculate the corresponding α-cuts of possibility distributions $\left(\pi_{\tilde{X}_{k+1}},\ldots,\pi_{\tilde{X}_n}\right)$ as the intervals of the possibilistic variables $\left(\tilde{X}_{k+1},\ldots,\tilde{X}_n\right)$.

2.2 Compute the minimal and maximal values of the outputs of the model $f\left(X_1,\ldots,X_k,\tilde{X}_{k+1},\ldots,\tilde{X}_n\right)$, denoted by $\underline{f}_\alpha^i$ and $\overline{f}_\alpha^i$, respectively. In this computation, the probabilistic variables are fixed at the sampled values $\left(x_1^i,x_2^i,\ldots,x_k^i\right)$ whereas the possibilistic variables take all values within the ranges of the α-cuts of their possibility distributions $\left(\pi_{\tilde{X}_{k+1}},\ldots,\pi_{\tilde{X}_n}\right)$.

2.3 Record the extreme values $\underline{f}_\alpha^i$ and $\overline{f}_\alpha^i$ as the lower and upper limits of the α-cuts of $f\left(x_1^i,x_2^i,\ldots,x_k^i,\tilde{X}_{k+1},\ldots,\tilde{X}_n\right)$.

End

3) Collect all the lower and upper limits of the different α-cuts of $f\left(x_1^i,x_2^i,\ldots,x_k^i,\tilde{X}_{k+1},\ldots,\tilde{X}_n\right)$ to establish an approximated possibility distribution (denoted by π_i^f) of the model output.

End

This procedure results in an ensemble of m realizations of the approximated possibility distributions $\left(\pi_1^f,\ldots,\pi_m^f\right)$. For each set A in the universe of discourse of all power adequacy values, the following formulas are used to obtain the possibility measure $Pos_i^f\left(A\right)$ and the necessity measure $Nec_i^f\left(A\right)$, given the possibility distribution π_i^f:

$$Pos_i^f\left(A\right)=\sup_{\{x\in A\}}\left\{\pi_i^f\left(x\right)\right\},$$

$$Nec_i^f\left(A\right)=\inf_{\{x\in A\}}\left\{1-\pi_i^f\left(x\right)\right\}. \tag{7.6}$$

These m different possibility and necessity measures are, then, used to obtain the belief $Bel\left(A\right)$ and the plausibility $Pl\left(A\right)$ of any set A, respectively:

$$Pl\left(A\right)=\sum_{i=1}^{m}p_i Pos_i^f\left(A\right),$$

$$Bel\left(A\right)=\sum_{i=1}^{m}p_i Nec_i^f\left(A\right), \tag{7.7}$$

where p_i is the probability of sampling the i-th realization $(x_1^i,x_2^i,\ldots,x_k^i)$ of the random variable vector $(X_1,\ldots,X_k)$. For each set A, this algorithm computes the probability-weighted average of the possibility measures associated with each output fuzzy interval.

For pure probabilistic propagation, the possibilistic distributions have to be converted into pdfs. This conversion can be achieved by various techniques [10], e.g. by simple normalization:

$$p_{X_i}(x) = \frac{\pi_{\tilde{X}_i}(x)}{\int_0^{+\infty} \pi_{\tilde{X}_i}(x)dx}. \tag{7.8}$$

Once the probabilistic distribution for each fuzzy variable is determined, the outer loop of the algorithm is performed m times, and at each iteration, the vector $(X_1, X_2, \ldots, X_n)$ is sampled and the corresponding adequacy value is calculated. After the m repetitions, the empirical probability distribution of system adequacy P_A is obtained.

7.1.4 Case Study

The system used as case study is modified from the IEEE 34 node distribution test feeder. Detailed information about this study can be found in [1]. Figures 7.2–7.4 present the graphical comparisons between the empirical cumulative distribution function (cdf) obtained by the probabilistic propagation approach and the belief and plausibility functions obtained by the joint propagation approach at different renewable penetration levels. The following observations can be drawn from the comparisons:

1) The cdf of distributed generators adequacy obtained by the pure probabilistic approach lies within the boundaries of belief and plausibility functions obtained by the joint propagation approach.
2) An explicit separation exists between the belief and plausibility functions reflecting the total imprecision of the information concerning the renewable generators parameters.
3) The separation between belief function and plausibility function grows with the penetration level, yet the empirical cdf remains relatively stable. More detailed analysis of the results and discussions on their implementation can be found in [1].

7.2 Nuclear Power Plant Components Degradation

7.2.1 Dissimilar Metal Weld Degradation

The cracking process in an Alloy 82/182 dissimilar metal weld in a primary coolant system can follow three major morphologies [11]: axial, radial, and circumferential. The latter two types can lead to the rupture of the component. The crack growth has two steps: crack initiation and crack propagation. The radial crack mainly grows outward from the initiation site toward the outer diameter; the process can lead to a leak and potentially to rupture. The crack grows evenly around the circumference, potentially leading to rupture.

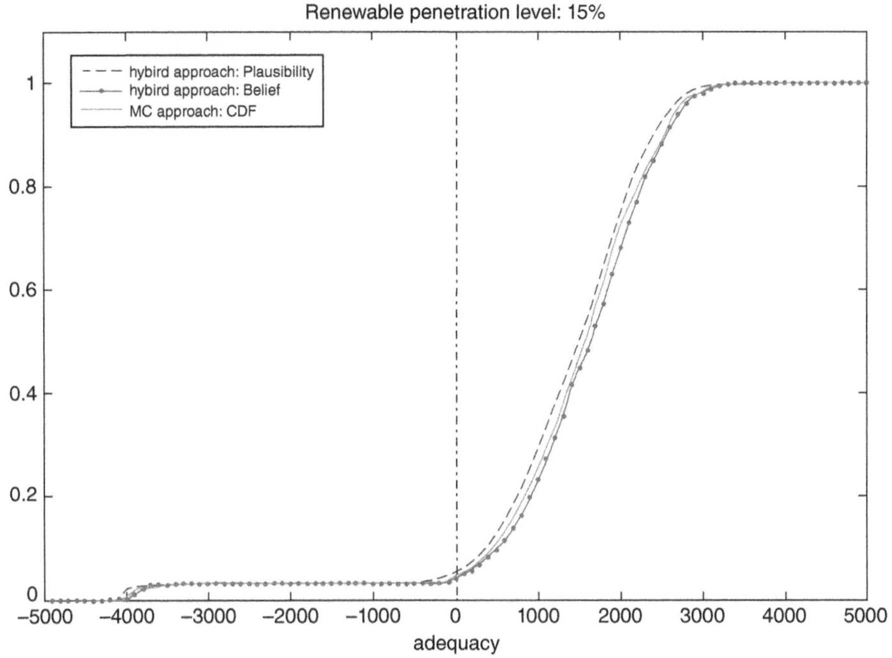

Figure 7.2 Comparison of joint propagation and pure probabilistic approaches at a renewable penetration level of 15% [1].

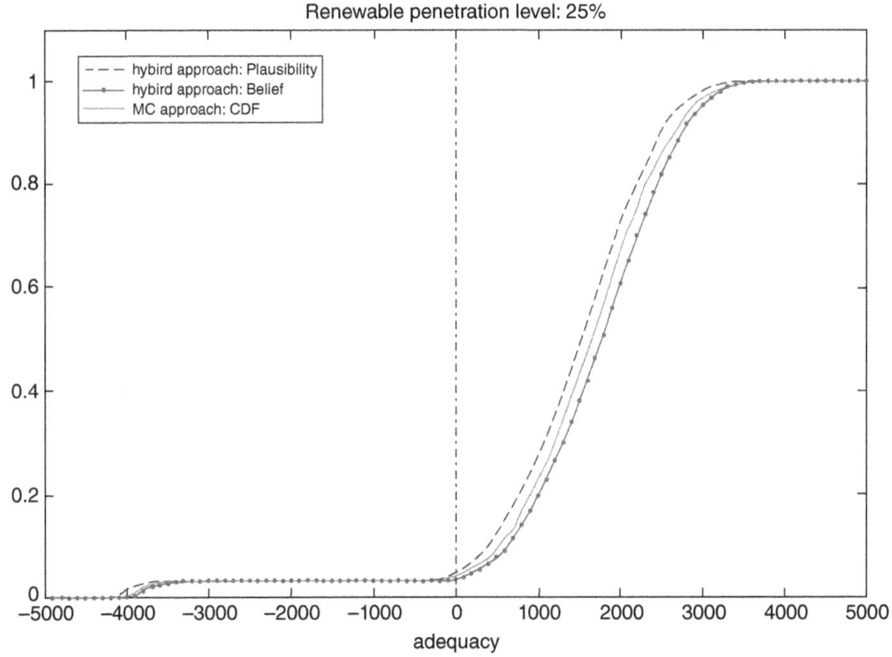

Figure 7.3 Comparison of joint propagation and pure probabilistic approaches at a renewable penetration level of 25% [1].

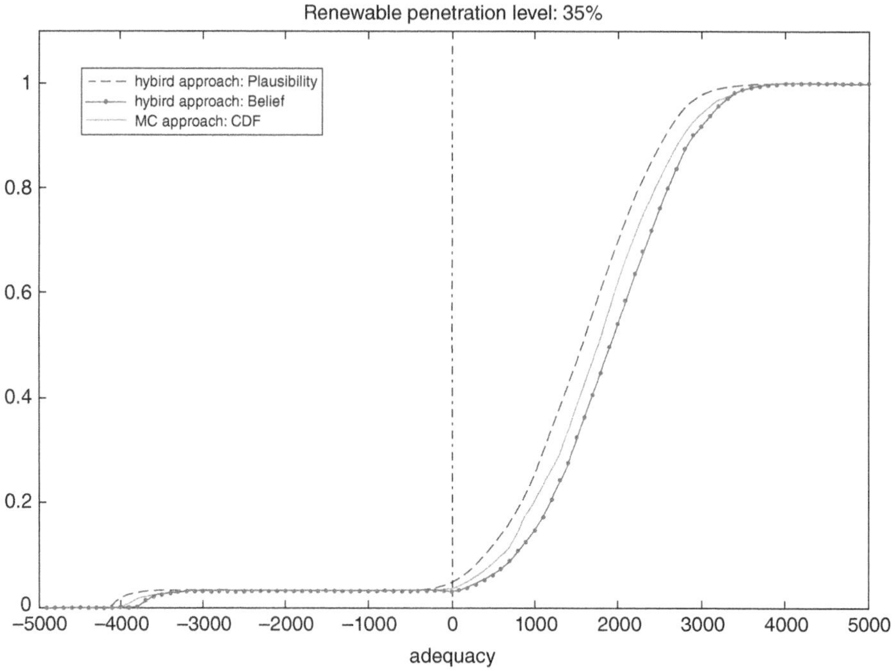

Figure 7.4 Comparison of joint propagation and pure probabilistic approaches at a renewable penetration level of 35% [1].

The Alloy 82/182 crack growth rate equations have been studied by various organizations including Ringhals AB, Electricité de France, and the Electric Power Research Institute. These equations take a similar form and include a stress and Arrhenius temperature dependence:

$$\dot{a} = \frac{da}{dt} = \alpha f_{alloy} f_{orient} K^{\beta} e^{\left[-(Q/R)\left(1/T - 1/T_{ref}\right)\right]} \tag{7.9}$$

where $\dot{a}$ ($\dot{a} \geq 0$) is the crack growth rate in time, a is the crack length (m), t is the time since crack initiation (s), α is the crack growth amplitude, f_{alloy} is a constant (equal to 1.0 for Alloy 182 and 0.385 for Alloy 82), f_{orient} is a constant equal to 1.0, K is the crack tip stress intensity factor (MPa$\sqrt{}$m), β is the stress intensity exponent, Q is the thermal activation energy for crack growth (kJ/mole), R is the universal gas constant (kJ/mole-^{o}K), T is the absolute operating temperature at crack location (^{o}K), and T_{ref} is the absolute reference temperature used to normalize data (^{o}K).

The multi-state physics model, proposed by Unwin, et al. [11] to describe the crack growth in the case study, is represented in Figure 7.5.

In [11], the transition rates φ_1, φ_2,φ_3 , and φ_4 are time-dependent and stochastic; the others are assumed constant.

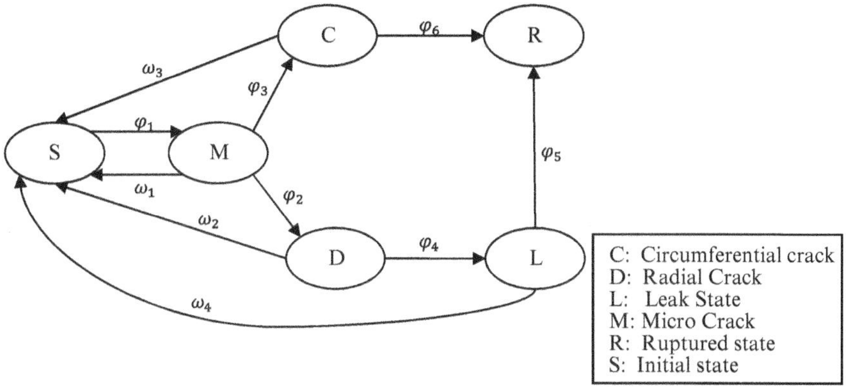

Figure 7.5 Transition diagram of the multi-state physics model of crack development in Alloy 82/182 dissimilar metal welds [2].

The transition rate φ_1 from initial state S to micro-crack state M is defined as

$$\varphi_1 = \int \left(\frac{b}{\tau}\right) \cdot \left(\frac{t}{\tau}\right)^{b-1} \cdot f_{PDF}\left(\tau, b\right) d\tau db, \tag{7.10}$$

where $f_{PDF}\left(\tau, b\right)$ is the joint probability density function of τ and b, and the integral is defined on the domains of τ and b. The parameter τ is a time constant, which has been observed to have a stress and temperature dependence; b is a fitting parameter.

The transition rates φ_2 and φ_3 describing the transitions from micro-crack state M to radial-crack state D and circumferential-crack state C, respectively, have similar definitions. Let a_D denote the threshold length of a radial-crack; then, at time u after crack initiation, the probability of the state D is defined as:

$$D\left(u\right) = P_D \cdot Pr\left[a_D \leq \int_0^u \dot{a}\left(t\right) dt\right], \tag{7.11}$$

where P_D is the probability that the crack grows to state D with the current state is M. The analogous probability $C\left(u\right)$ that the crack goes to state C at time u after crack initiation is defined as:

$$C\left(u\right) = P_C \cdot Pr\left[a_C \leq \int_0^u \dot{a}\left(t\right) dt\right], \tag{7.12}$$

where a_C is the threshold length of a circumferential crack and P_C is the probability that the crack goes to state C given that the current state is M.

The transition rate φ_2 (between state M and D) is defined as [12]:

$$\varphi_2 = \frac{dD(u)/du}{1-D(u)} = \frac{(a_D/u)^2 \cdot \pi(a_D/u)P_D}{1-P_D\int_{a_D/u}^{\infty}\pi(\dot{a})d\dot{a}}, \tag{7.13}$$

and similarly

$$\varphi_3 = \frac{dC(u)/du}{1-C(u)} = \frac{(a_C/u)^2 \cdot \pi(a_C/u)P_C}{1-P_C\int_{a_C/u}^{\infty}\pi(\dot{a})d\dot{a}}, \tag{7.14}$$

By assuming the crack growth rate $\dot{a}$ is following a uniform distribution with a maximum value of $\dot{a}_M$, i.e.:

$$\pi(\dot{a}) = \begin{cases} \frac{1}{\dot{a}_m}, & if\, 0 < \dot{a} < \dot{a}_m \\ 0, & else \end{cases}, \tag{7.15}$$

then, Equations (7.13) and (7.14) are reduced to:

$$\varphi_2 = \begin{cases} \frac{a_D P_D}{\dot{a}_M u^2\left(1-P_D\left(1-a_D/(u\dot{a}_M)\right)\right)}, & if\, u > a_D/\dot{a}_M \\ 0, & otherwise \end{cases}, \tag{7.16}$$

and

$$\varphi_3 = \begin{cases} \frac{a_C P_C}{\dot{a}_M u^2\left(1-P_C\left(1-a_C/(u\dot{a}_M)\right)\right)}, & if\, u > a_C/\dot{a}_M \\ 0, & else \end{cases}, \tag{7.17}$$

respectively.

The transition rate φ_4 from state D to state L is defined by the growth in crack size up to a threshold a_L of leakage:

$$L(w) = Pr\left[a_L - a_D \leq \int_0^w \dot{a}(t)dt\right], \tag{7.18}$$

$$\varphi_4 = \frac{dL(w)/dw}{1-L(w)}, \tag{7.19}$$

where w is the time from the radial crack formation [21]. By assuming the same distribution over the crack growth rate, then

$$\varphi_4 = \begin{cases} \frac{1}{w}, & if\ w > (a_L - a_D) / \dot{a}_M \\ 0, & otherwise \end{cases} . \tag{7.20}$$

Transition rates from leak to rupture and from circumferential crack to rupture are assumed to be constant. These transition rates, together with other constant parameters, are presented in Table 7.2.

7.2.2 MCS Method

The multi-state physical models, like that presented in Figure 7.5, are non-homogenous Markov processes and typically have no closed-form solutions. Thus, simulation algorithms or numerical methods are developed to solve them. In this section, we will introduce the developed simulation procedures for the model considered. For the details about the theoretical foundation of this method, please refer to [2].

Prior to the simulation, external influencing factors should be incorporated through the following three steps.

1) Formulate the functions describing the physics of the transition rates.
2) Identify the external influencing factors θ_i (e.g. temperature, stress).
3) Define the distribution functions, $p(\theta)$, representing the uncertainties in the values of these factors.

The algorithm for simulating the component degradation process on the time horizon $[0, t_{max}]$ is sketched in the following pseudocode.

Table 7.2 Case study parameter definitions and values [2].

b -Weibull shape parameter for crack initiation model	2.0
τ - Weibull scale parameter for crack initiation model	4 years
a_D - Crack length threshold for radial macro-crack	10 mm
P_D - Probability that micro-crack evolves as radial crack	0.009
$\dot{a}_M$ - Maximum credible crack growth rate	9.46 mm/yr
a_C - Crack length threshold for circumferential macro-crack	10 mm
P_C - Probability that micro-crack evolves as circumferential crack	0.001
a_L - Crack length threshold for leak	20 mm
ω_1 - Repair transition rate from micro-crack	$1 \times 10^{(-3)}$ /yr
ω_2 - Repair transition rate from radial macro-crack	$2 \times 10^{(-2)}$ /yr
ω_3 - Repair transition rate from circumferential macro-crack	$2 \times 10^{(-2)}$ /yr
ω_4 - Repair transition rate from leak	$8 \times 10^{(-1)}$ /yr
φ_5 - Leak to rupture transition rate	$2 \times 10^{(-2)}$ /yr
φ_6 - Macro-crack to rupture transition rate	$1 \times 10^{(-5)}$ /yr

Initialize the system by allocating a token onto place $i = M$ (initial state of perfect performance), setting the time $t = 0$ (initial time) and the total number of replications to N_{max}.

Set $t' = 0$.

Set $n = 1$.

While $n < N_{max}$,

While $t < t_{max}$,

Sample a realization of the external influencing factors $\boldsymbol{\theta}$ from the joint probability function $p(\boldsymbol{\theta})$.

Sample a departure time t from the distribution function $F_i(t \mid t', \boldsymbol{\theta})$.

Sample a random number U from the uniform distribution in [0, 1].

For each outgoing transition ($j = 0, 1, ..., M, j \neq i$),

Calculate the transition probability $q_{i,j}(t, \boldsymbol{\theta})$.

If $\sum_{k=0}^{j^*-1} q_{i,k} < U < \sum_{k=0}^{j^*} q_{i,k}$,

then activate the transition to state j^*.

End If.

End For.

Set $t' = t$.

Remove the token from place i and add a new token onto place j^*.

End While.

Set $n = n + 1$.

End While.

Subsequent to the execution of the simulation algorithm, an estimate $\hat{P}(t) = \left\{\hat{P}_0(t), \hat{P}_1(t), \ldots, \hat{P}_M(t)\right\}$ of the state probability vector is computed by dividing the total number of visits to each state by the total number of simulations N_R: $\hat{P}(t) = \frac{1}{N_{max}}\{n_0(t), n_1(t), \ldots, n_M(t)\}$, where $\{n_i(t) \mid i = 0, \ldots, M,\ t \leq t_{max}\}$ is the total number of visits to state i at time t. The distributions $p(\boldsymbol{\theta})$ and $F_i(t \mid t')$ may have complicated mathematical expressions; under these circumstances, the Markov Chain Monte Carlo (MCMC) technique can be used to sample random values [13]. There are two key quantities in the simulation procedure above: $F_i(t \mid t')$ and $q_{i,j}(t, \boldsymbol{\theta})$. The former is the cdf of the departure time t given that it is at state i at time t' and is defined as:

$$F_i(t \mid t') = \int F_i(t \mid t', \boldsymbol{\theta}) p(\boldsymbol{\theta}) d\boldsymbol{\theta} = 1 - \int \exp\left[-\int_{t'}^{t} \lambda_i(t'', \boldsymbol{\theta}) dt''\right] p(\boldsymbol{\theta}) d\boldsymbol{\theta}. \tag{7.21}$$

The latter is the marginal transition probability to any other state $j = \{0,\ldots M \mid j \neq i\}$ given that the present state is i. It is defined as:

$$q_{i,j}(t) = \int \frac{\lambda_{i,j}(t,\boldsymbol{\theta})}{\lambda_i(t,\boldsymbol{\theta})} p(\boldsymbol{\theta}) d\boldsymbol{\theta}. \tag{7.22}$$

The details about the computation of these quantities can be found in [2].

7.2.3 Numerical Results

The simulation model has been executed $N_{max} = 10^7$ times over a component lifetime $t_{max} = 80$ years in line with the original study. To investigate the convergence of the simulation model, the 10^7 realizations have been subdivided into $N = 20$ subsamples of 500,000 each. The sample mean and variance of the estimated state probabilities are calculated as

$$\bar{\boldsymbol{P}}(t) = \frac{1}{N}\sum_{k=1}^{N} \hat{\boldsymbol{P}}(t)_k, \tag{7.23}$$

$$var_{\hat{\boldsymbol{P}}(t)} = \frac{1}{N-1}\sum_{k=1}^{N}\left[\hat{\boldsymbol{P}}(t)_k - \bar{\boldsymbol{P}}(t)\right]^2, \tag{7.24}$$

where $\hat{\boldsymbol{P}}(t)_k$ is the estimated state probability vector from the k-th subsample. The convergence of the state probability values can be observed by the variance in (7.24) and the sequence of sample means on the steadily incremental subsamples, i.e.

$$\boldsymbol{P}(t)_{conv,k} = \frac{1}{n}\sum_{k=1}^{n} \hat{\boldsymbol{P}}(t)_k, \tag{7.25}$$

where n takes value from 1 to N.

At $t = 80$ years, the variances are 0.6749×10^{-8}, 0.776×10^{-8}, 0.0352×10^{-8}, 0.0106×10^{-8}, 0.0037×10^{-8}, and 0.0337×10^{-8} for initial, micro-crack, circumferential, radial, leak, and rupture states, respectively. Similar results are found at different time moments. The examples of convergence curves at 80 years are presented in Figure 7.6. The good stabilization of $\boldsymbol{P}(80)_{conv,k}$ is manifested. $\boldsymbol{P}(t)_{conv,k} = \bar{\boldsymbol{P}}(t)$. Similar convergence curves are obtained at different time moments but are not presented to save space.

The numerical comparisons on the state probability values at year 80 are reported in Table 7.3. As expected, the relative differences (i.e. the differences between the state probability values computed by the simulation method minus those obtained with the state-space enrichment method, divided by the former) decrease as the step size is reduced. For the details about the state-space enrichment method, please refer to [14].

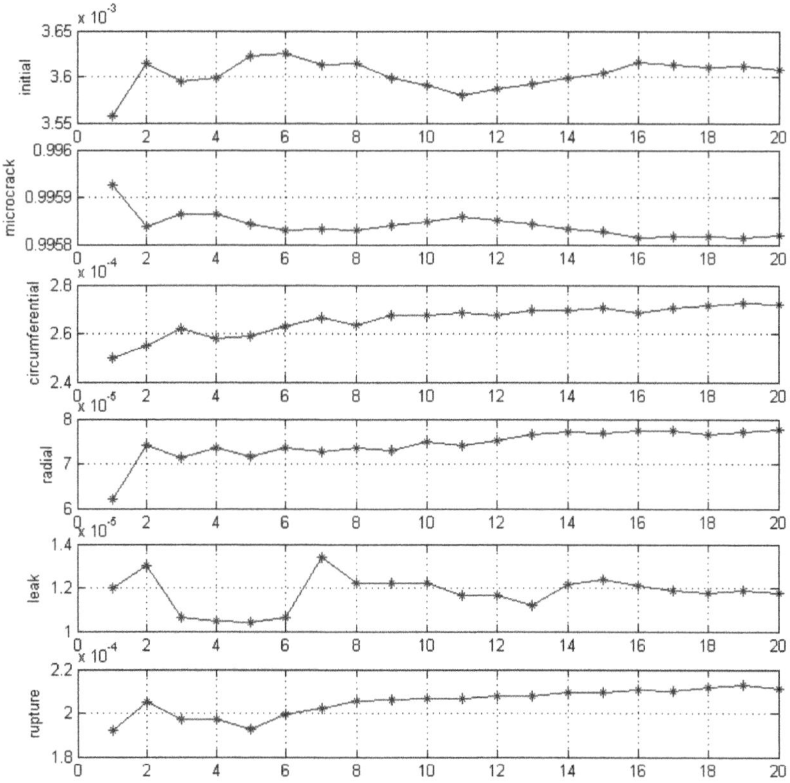

Figure 7.6 Convergence plots of state probabilities at t = 80 years [2].

Table 7.3 Comparison of the simulation results with the state-space enrichment results (state probability values at year 80) [2].

	Simulation	**State-space enrichment Step size = 1 year**	**Relative difference**	**State-space enrichment Step size = 0.5 year**	**Relative difference**	**State-space enrichment Step size = 0.1 year**	**Relative difference**
Initial state probability	*0.0036*	0.0033	8.33%	0.0034	5.56%	0.0036	0.00%
Micro-crack probability	*0.9958*	0.9963	-0.05%	0.9961	-0.03%	0.9959	-0.01%
Circumferential crack probability	*2.72e-4*	1.94e-04	28.68%	2.33e-04	14.34%	2.78e-04	-2.21%
Radial crack probability	*7.78e-5*	6.38e-05	17.99%	6.97e-05	10.41%	7.66e-05	1.54%
Leak probability	*1.18e-5*	8.93e-06	24.32%	1.06e-05	10.17%	1.24e-05	-5.08%
Rupture state probability	*2.11e-4*	1.38e-04	34.60%	1.73e-04	18.01%	2.12e-04	-0.47%

References

1 Li, Y. and Zio, E. (2012). Uncertainty analysis of the adequacy assessment model of a distributed generation system. *Renewable Energy* 41: 235–244.

2 Li, Y.-F., Zio, E., and Lin, Y.-H. (2012). A multistate physics model of component degradation based on stochastic petri nets and simulation. *IEEE Transactions on Reliability* 61 (4): 921–931.

3 Hegazy, Y., Salama, M., and Chikhani, A. (2003). Adequacy assessment of distributed generation systems using Monte Carlo simulation. *IEEE Transactions on Power Systems* 18 (1): 48–52.

4 Karki, R., Hu, P., and Billinton, R. (2010). Reliability evaluation considering wind and hydro power coordination. *IEEE Transactions on Power Systems* 25 (2): 685–693.

5 El-Khattam, W., Hegazy, Y., and Salama, M. (2006). Investigating distributed generation systems performance using Monte Carlo simulation. *IEEE Transactions on Power Systems* 21 (2): 524–532.

6 Allan, R.N. (2013). *Reliability Evaluation of Power Systems*. Chester, England: Springer Science & Business Media.

7 Conti, S. and Raiti, S. (2007). Probabilistic load flow using Monte Carlo techniques for distribution networks with photovoltaic generators. *Solar Energy* 81 (12): 1473–1481.

8 Baudrit, C., Dubois, D., and Guyonnet, D. (2006). Joint propagation and exploitation of probabilistic and possibilistic information in risk assessment. *IEEE Transactions on Fuzzy Systems* 14 (5): 593–608.

9 Baraldi, P. and Zio, E. (2008). A combined Monte Carlo and possibilistic approach to uncertainty propagation in event tree analysis. *Risk Analysis* 28 (5): 1309–1326.

10 Flage, R., Aven, T., and Zio, E., "Alternative representations of uncertainty in system reliability and risk analysis: Review and discussion," in *ESREL 2008*, 2008, pp. 2081–2091.

11 Unwin, S.D., Lowry, P.P., Layton, R.F., Heasler, P.G., and Toloczko, M.B. (2011). Multi-state physics models of aging passive components in probabilistic risk assessment. In: *Proceedings of ANS PSA 2011 International Topical Meeting on Probabilistic Safety Assessment and Analysis*, 1–12.

12 Fleming, K.N. et al. (2010). Treatment of passive component reliability in risk-informed safety margin characterization. Idaho National Laboratory, Idaho Falls, Idaho, INL/EXT-10-20013.

13 Zio, E. and Zoia, A. (2009). Parameter identification in degradation modeling by Reversible-Jump Markov Chain Monte Carlo. *IEEE Transactions on Reliability* 58 (1): 123–131.

14 Unwin, S.D., Lowry, P.P., Layton, R.F., Heasler, P.G., and Toloczko, M.B. (2011). Multi-state physics models of aging passive components in probabilistic risk assessment. In: *ANS PSA 2011 International Topical Meeting on Probabilistic Safety Assessment and Analysis*, 1–12. Richland, WA: Pacific Northwest National Laboratory (PNNL).

Part III

Optimization Methods and Applications

Reliability optimization aims at maximizing system reliability and related metrics, while minimizing the cost associated to the reliability improvements. It has been an active research domain since the 1960s. Various reliability optimization problems have been formulated and various solution techniques proposed. In general, the decision variables of such optimization problems encode the parameters driving the system reliability improvements, for example the inherent component reliability (and the related parameters, like failure probability, failure rate, etc.), the system logic configuration (e.g. the number of subsystems in series, the number of redundant components, the components type, etc.), which determine the system reliability allocation, and those relevant to testing and maintenance activities (e.g. the test intervals, maintenance periodicities, etc.), which determine the system availability and maintainability characteristics. The generic formulation of the reliability optimization problem typically consists of two parts: the objective function (of the decision variables), which is defined so as to lead to maximize reliability (or minimize unavailability); the constraints, which ensure that the resources, e.g. cost and weight, used to enhance system reliability are under certain limits. This is not the only formulation, as the objective function and constraints are interchangeable and can be combined in multi-objective formulations. For example, in some formulations the objective function is cost minimization and there is one constraint requiring a reliability value higher than a certain level.

The objective and constraints are mostly nonlinear and the decision variables can be a mix of continuous variables (e.g. test interval) and integer variables (e.g. number of redundant components). In addition, uncertainties could exist in the parameters of the objective function and constraints, and in the coefficients of the problem. These characteristics render the reliability optimization problems generally quite difficult to solve. Solution techniques applied to reliability optimization problems have been well documented in the surveys by Kuo and Prasad [1] and by Kuo and Wan [2].

This part of the book collects various formulations and solution methods, covering from the conventional mathematical programming to the latest robust optimization methods. In Chapter 8, the mathematical programming approaches are introduced. In Chapter 9, evolutionary algorithms are presented. Since reliability optimization is essentially multi-objective, in Chapter 10 the multi-objective formulation and the solution methods are introduced. Chapter 11 presents the optimization under uncertainty and Chapter 13 presents applications and case studies.

8

Mathematical Programming

Mathematical programming methods, e.g. linear programming, integer programming (IP), convex optimization, aim to search for the exact solution(s) to an optimization problem in tractable time. They have very good theoretical foundations as well as a long history in solving various practical and challenging optimization problems with success. However, reliability optimization problems are typically non-linear and complex in nature, thus meta-heuristic and evolutionary algorithms are being often adopted because they are relatively straightforward to comprehend and implement.

The major drawback of evolutionary algorithms is the inability to guarantee the global optimal solution. On the contrary, mathematical programming methods can achieve the global optimal solution though they might be relatively difficult to comprehend and time-consuming to perform in some cases. Yet when addressing a reliability optimization problem, one should first analyze its mathematical properties and consider whether a mathematical programming method is suitable for solving the problem. Evolutionary algorithms can be considered if the mathematical programming methods are infeasible or inefficient. Thus, as the first chapter of the optimization part of this book, it is devoted to two basic mathematical programming methods. For more advanced knowledge about mathematical programming, please refer to the books [1-3].

8.1 Linear Programming (LP)

In reliability optimization problems, *component reliability enhancement* is among the most important ones. The problem amounts to optimizing the reliability of the system by improving the reliability of its components. Already in 1973, Kulshrestha and Gupta [4] had formulated one such problem to maximize the reliability of a series system as follows:

$$\max_{r_i, i=1,2,\ldots,n} R = \prod_{i=1}^{n} r_i \tag{8.1a}$$

Reliability Analysis, Safety Assessment and Optimization: Methods and Applications in Energy Systems and Other Applications, First Edition. Enrico Zio and Yan-Fu Li.

$$\text{s.t.} \sum_{i=1}^{n} h_{ji}(r_i) \leq b_j,\ j = 1,2,\ldots,m \tag{8.1b}$$

$$\alpha_i \leq r_i \leq \beta_i, \forall i \tag{8.1c}$$

$$0 < \alpha_i < \beta_i < 1, \forall i \tag{8.1d}$$

where n is the number of subsystems (i.e. stages), r_i is the reliability of subsystem i (for example, it corresponds to a simple component i, considering a series system that has only one component in each subsystem), $h_{ji}(r_i)$ is the j-th resource consumed for subsystem i, and b_j is the total amount of resource j available, $j = 1,2,\ldots,m$. This problem is also referred to as the system reliability allocation, and it is among the earliest attempts to system reliability optimization.

Assuming a resource function of the form $h_{ji}(r_i) = M_{ji} + c_{ji}\ln(r_i)$ where $M_{ji} \geq -c_{ji}\ln(\alpha_i)$, taking the logarithm of the objective function, and letting $\ln(r_i) = y_i$, we have the following converted problem:

$$\max_{y_i, i=1,2,\ldots,n} \ \ln R = \sum_{i=1}^{n} y_i \tag{8.2a}$$

$$\text{s.t.} \sum_{i=1}^{n} \left(M_{ji} + c_j y_i\right) \leq b_j,\ j = 1,2,\ldots,m \tag{8.2b}$$

$$\ln(\alpha_i) \leq y_i \leq \ln(\beta_i), \forall i \tag{8.2c}$$

$$0 < \alpha_i < \beta_i < 1, \forall i \tag{8.2d}$$

In this formulation, the objective function and all constraints are linear, and the decision variables are continuous. This is referred to as a *linear programming* (LP) problem which is the foundation of other mathematical programming problems, like linear integer programming (IP).

Example 8.1 Suppose two subsystems are in a series system, and their reliabilities are r_1 and r_2, respectively. The converted variables are y_1 and y_2, respectively. There are two constraints and the limits of y_1 and y_2 are [-0.7, -0.001]. The problem formulation of this example is shown as follows:

$$\max\ y_1 + y_2 \tag{8.3a}$$

$$\text{s.t. } 70 + 100 y_1 + 35 + 50 y_2 \leq 95 \tag{8.3b}$$

$$50 + 50 y_1 + 140 + 200 y_2 \leq 180 \tag{8.3c}$$

$$-0.7 \leq y_i \leq -0.001, \forall i \tag{8.3d}$$

Because only two variables exist, we can solve the problem by representing the set of points that satisfy all constraints (i.e. feasible region) on the two-dimensional plane and searching for the point that maximizes the objective function. Each inequality constraint is satisfied by a half-plane of points and the feasible region is the intersection of all these half-planes. Figure 8.1 illustrates this process of the graphical solution method. The shaded area is the feasible region of the LP problem and the dashed line is the objective to be maximized, i.e. $y_1 + y_2$. As one moves the line from the bottom-left corner up and to the right, then the value of $y_1 + y_2$ increases. Thus, we look for the line that is furthest from the bottom-left corner and still touches the feasible region. This occurs at the intersection of the lines $10y_1 + 5y_2 = -1$ and $5y_1 + 20y_2 = -1$. Thus, the optimal solution of the problem is (-3/35, -1/35), and the corresponding maximal value of the objective function is -4/35.

8.1.1 Standard Form and Duality

In general, all LP problems can be converted into a standard form where all decision variables are non-negative and the main constraints are inequalities. Let $\boldsymbol{x} = \left(x_1, \ldots, x_n\right)^T$ denote the vector of the decision variables, $\boldsymbol{c} = \left(c_1, \ldots, c_n\right)^T$ denote the coefficients of the objective function, $\boldsymbol{b} = \left(b_1, \ldots, b_m\right)^T$ denote the right-hand side values of the inequality constraints, and

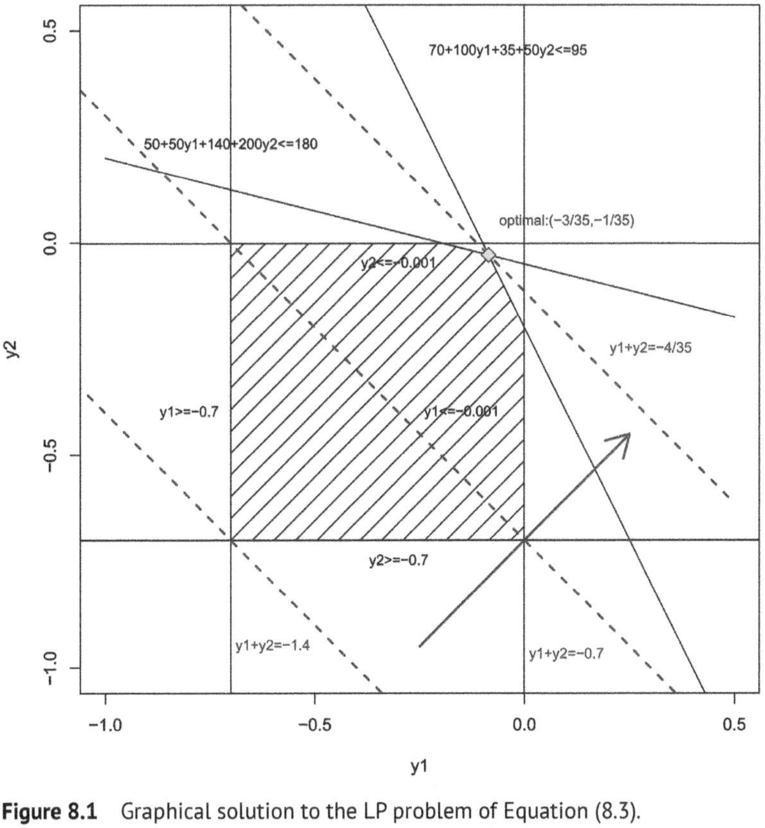

Figure 8.1 Graphical solution to the LP problem of Equation (8.3).

$$\boldsymbol{A} = \begin{pmatrix} a_{11} & \cdots & a_{1n} \\ \vdots & \ddots & \vdots \\ a_{m1} & \cdots & a_{mn} \end{pmatrix}$$

denote the coefficients of the constraints. The standard LP form in the case of maximization can be presented as follows:

$$\max_{x_i, i=1,2,\ldots,n} \sum_{i=1}^{n} c_i x_i \tag{8.4a}$$

$$\text{s.t.} \sum_{i=1}^{n} a_{ij} x_i \leq b_j,\ j = 1,\ldots,m \tag{8.4b}$$

$$x_i \geq 0,\ i = 1,\ldots,n \tag{8.4c}$$

where n is the number of decision variables and m is the number of the constraints. The above can also be written in the following matrix form:

$$\max \boldsymbol{c}^T \boldsymbol{x} \tag{8.5a}$$

$$\text{s.t.}\, \boldsymbol{Ax} \leq \boldsymbol{b} \tag{8.5b}$$

$$\boldsymbol{x} \geq 0 \tag{8.5c}$$

In case of an LP minimization problem, the objective becomes "min" and the inequalities become $\geq$. The standard LP problems are typically solved by the Simplex method [5], developed by G.B. Dantzig in 1947. The Simplex procedures are presented as follows:

- Determine the extreme points of the polygon (or polyhedron in higher-dimensional spaces) of the feasible region.
- Find some feasible extreme points and calculate the objective value, z.
- Test if an extreme point is optimal:
  - If no, move to the adjacent extreme point that gives the greatest rate of improvement in objective z, to perform the same test,
  - If yes, stop the process.
- Stop the process if an unbounded case occurs.

Example 8.2 In the following paragraphs, we will illustrate the Simplex method to solve the following LP problem:

$$\max_{x_1, x_2} 2x_1 + 3x_2 \tag{8.6a}$$

$$\text{s.t.}\, x_1 + x_2 \leq 20 \tag{8.6b}$$

$$0 \leq x_1 \leq 16, 0 \leq x_2 \leq 10 \tag{8.6c}$$

As seen in Figure 8.1, the optimal solution lies on one of the extreme points. Thus, the first step of the Simplex method is to find all the extreme points of the feasible region. To complete this task, we introduce three slack variables to convert inequality constraints to equality constraints. Then the original problem becomes the following:

$$\max_{x_1,x_2,s_1,s_2,s_3} \quad 2x_1+3x_2+0s_1+0s_2+0s_3 \tag{8.7a}$$

$$\text{s.t.}\, x_1+x_2+s_1=20 \tag{8.7b}$$

$$x_1+s_2=16 \tag{8.7c}$$

$$x_2+s_3=10 \tag{8.7d}$$

$$x_i \geq 0\left(i=1,2\right),\ s_i \geq 0\left(i=1,2,3\right) \tag{8.7e}$$

Now the task is to solve the linear model of Equations (8.7b–8.7d) with five variables. Because there are two equations less than variables, we assign two variables to be zeros and then solve the remaining 3×3 linear system. The unique solution to this system is a basic solution. Any basic feasible solution (BFS) corresponds to an extreme point where all its variables are non-negative.

The disadvantage is that the number of BFSs could be large. For example, suppose we have 500 variables and 400 equality constraints: This gives a number of BFSs of the order of $\begin{pmatrix}500\\100\end{pmatrix}$, and it is difficult or even impossible, to explore all BFSs to find the optimal extreme point.

Alternatively, we can start with some extreme points and move to the adjacent one which gives the highest rate of improvement in the objective function. This process iterates until we find the optimal extreme point. The canonical form of Equation (8.7) is illustrative to identify the extreme points and move from one point to another. Let $z=2x_1+3x_2+0s_1+0s_2+0s_3$, the canonical form is shown as follows:

$$\begin{pmatrix} 1 & -2 & -3 & 0 & 0 & 0 \\ 0 & 1 & 1 & 1 & 0 & 0 \\ 0 & 1 & 0 & 0 & 1 & 0 \\ 0 & 0 & 1 & 0 & 0 & 1 \end{pmatrix} \begin{pmatrix} z \\ x_1 \\ x_2 \\ s_1 \\ s_2 \\ s_3 \end{pmatrix} = \begin{pmatrix} 0 \\ 20 \\ 16 \\ 10 \end{pmatrix} \tag{8.8}$$

In this form, the dependent variable has only one non-zero entry in its corresponding column in the 4×6 matrix. In other words, the dependent variables are expressed by the independent variables. In Equations (8.7–8.8), the set of dependent variables is $\{s_1, s_2, s_3\}$ and the set of independent variables is $\{x_1, x_2\}$. Then, the process is presented as follows:

Step 1: Select one initial extreme point.
We assign the independent variables $x_1 = 0$ and $x_2 = 0$. In this case, we have $z = 0$.

Step 2: Select one dependent variable to enter the independent variable set and one independent variable to enter the dependent variable set.
By looking at the objective function, if we increase x_1 by one unit, we can increase z by two units; if we increase x_2 by one unit, then we can increase z by three units. Therefore, we let x_2 be the dependent variable.

On the other hand, we cannot increase x_2 without limit, but we want to increase it as much as possible. Holding $x_1 = 0$, we have $x_2 + s_1 = 20$, $s_2 = 16$ and $x_2 + s_3 = 10$. Then x_2 cannot surpass 10 and, thus, s_3 should be the independent variable because in this case $s_3 = 0$.

We need the set of dependent variables $\{s_1, s_2, x_2\}$ and the set of independent variables $\{x_1, s_3\}$. To achieve this, we have to transform the linear model. In canonical form, this change means that the column corresponding to x_2 needs to have only one non-zero entry value 1. The following operation is conducted:

$$
\begin{matrix} *3 \\ *(-1) \\ \\ \\ \end{matrix}
\begin{pmatrix} 1 & -2 & -3 & 0 & 0 & 0 \\ 0 & 1 & 1 & 1 & 0 & 0 \\ 0 & 1 & 0 & 0 & 1 & 0 \\ 0 & 0 & 1 & 0 & 0 & 1 \end{pmatrix}
\begin{pmatrix} z \\ x_1 \\ x_2 \\ s_1 \\ s_2 \\ s_3 \end{pmatrix} = \begin{pmatrix} 0 \\ 20 \\ 16 \\ 10 \end{pmatrix}
$$

$$
\Downarrow
$$

$$
\begin{pmatrix} 1 & -2 & 0 & 0 & 0 & 3 \\ 0 & 1 & 0 & 1 & 0 & -1 \\ 0 & 1 & 0 & 0 & 1 & 0 \\ 0 & 0 & 1 & 0 & 0 & 1 \end{pmatrix}
\begin{pmatrix} z \\ x_1 \\ x_2 \\ s_1 \\ s_2 \\ s_3 \end{pmatrix} = \begin{pmatrix} 30 \\ 10 \\ 16 \\ 10 \end{pmatrix}
\qquad (8.9)
$$

Step 3: Obtain a new solution and repeat Step 2.
Given the set of dependent variables $\{s_1, s_2, x_2\}$ and the set of independent variables $\{x_1, s_3\}$, we obtain that $z = 30$ according to the canonical form in Equation (8.9). Again, by looking at the objective function, if we increase x_1 by one unit, then we increase z by two units; if we increase s_3 by one unit, we decrease z by three units. Therefore, we let x_1 be the dependent variable.

On the other hand, by holding $s_3 = 0$, we have $x_1 + s_1 = 10$, $x_1 + s_2 = 16$, and $x_2 = 10$. Then x_1 cannot surpass 10 and, thus, s_1 should be the independent variable.

Now we get the set of dependent variables $\{x_1, s_2, x_2\}$ and the set of independent variables as $\{s_1, s_3\}$. To achieve this, the following operation is conducted in the canonical form:

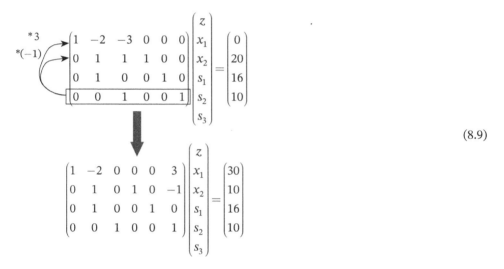

$$\begin{array}{l} \begin{array}{r} {}^{*}2 \\ {}^{*}(-1) \end{array} \begin{pmatrix} 1 & -2 & 0 & 0 & 0 & 3 \\ \boxed{0} & 1 & 0 & 1 & 0 & -1 \\ 0 & 1 & 0 & 0 & 1 & 0 \\ 0 & 0 & 1 & 0 & 0 & 1 \end{pmatrix} \begin{pmatrix} z \\ x_1 \\ x_2 \\ s_1 \\ s_2 \\ s_3 \end{pmatrix} = \begin{pmatrix} 30 \\ 10 \\ 16 \\ 10 \end{pmatrix} \\ \Downarrow \\ \begin{pmatrix} 1 & 0 & 0 & 2 & 0 & 1 \\ 0 & 1 & 0 & 1 & 0 & -1 \\ 0 & 0 & 0 & -1 & 1 & 1 \\ 0 & 0 & 1 & 0 & 0 & 1 \end{pmatrix} \begin{pmatrix} z \\ x_1 \\ x_2 \\ s_1 \\ s_2 \\ s_3 \end{pmatrix} = \begin{pmatrix} 50 \\ 10 \\ 6 \\ 10 \end{pmatrix} \end{array} \tag{8.10}$$

Step 4: Obtain a new solution and repeat Step 2.
Given the set of dependent variables $\{x_1, s_2, x_2\}$ and the set of independent variables $\{s_1, s_3\}$, we obtain that $z = 50$ according to the canonical form in Equation (8.10). Then we repeat Step 2 and we find that the increase of any variable in $\{s_1, s_3\}$ decreases z.

Step 5: Stop the iteration and output the results.
The optimal solution is $z = 50$ at $x_1 = x_2 = 10$.

In the above example, we have illustrated the Simplex method performed by hand. In practice, the problems are much larger in terms of the number of decision variables and constraints. A computer software must be used to solve them. For this, several automatized tools, e.g. CPLEX, LINDO, and XPRESS can efficiently solve LP problems.

8.2 Integer Programming (IP)

Redundancy allocation, first formulated by Ghare and Taylor [6] in 1969 and Beraha and Misra [7] in 1974, is a well-established way to reliability optimization. It aims to improve system reliability via installing redundant components into the system. Let y_{ij} denote the number of components (integer value) of the j-th version at the i-th subsystem. The formulation of the redundancy allocation problem (RAP) aims to maximize the total system reliability while keeping the system cost C not larger than a predefined level C_0. The formulation for the representative binary state series-parallel system (BSSPS) is presented as follows:

$$\max R(\boldsymbol{x}) = \prod_{i=1}^{N} \left(1 - \prod_{j=1}^{v_i} \left(1 - r_{ij}\right)^{x_{ij}} \right) \tag{8.11a}$$

$$\text{s.t.} \sum_{i=1}^{N} \sum_{j=1}^{v_i} c_{ij} x_{ij} \le C_0 \tag{8.11b}$$

$$u_{ij} \ge x_{ij} \ge l_{ij},\ x_{ij} \in \mathbb{Z}^* \tag{8.11c}$$

where v_i is the number of component versions available to the ith subsystem, r_{ij} is the reliability of the jth version component at the ith subsystem, $\boldsymbol{x}=\left(x_{11},\ldots,x_{1v_1};\ldots;x_{N1},\ldots,x_{Nv_N}\right)$ is the decision vector, and u_{ij} and l_{ij} are, respectively, the upper and lower limits of the number of jth version components at the ith subsystem.

The formulation in Equation (8.11) can be generalized as follows:

$$\min f(x) \tag{8.12a}$$

$$\text{s.t.}\, g(\boldsymbol{x})\geq \boldsymbol{b} \tag{8.12b}$$

$$\boldsymbol{x}\in\mathbb{Z}^n, \boldsymbol{x}\geq 0 \tag{8.12c}$$

where n is the number of non-negative integer decision variables. This type of problem is called an IP problem. In many cases, the IP problem is referred to as the integer linear programming (ILP) problem where the objective function and constraints (except the integer constraints) are all linear.

The ILP problem has a simple formulation, but in general, it is NP-complete, which means it is among the most difficult decision problems to solve. IP has been an active research area for decades. Several methods have been developed, and they can be mainly grouped into two classes: exact methods and heuristics. The exact methods, e.g. branch-and-bound (B&B) and cutting plane techniques, can obtain the global optimal solution, but they are restrained to certain problem types, e.g. ILP problems. The heuristics, e.g. hill climbing and genetic algorithms can work on any IP problems, but they cannot guarantee to obtain the global optimal solution.

In the following, we introduce B&B method procedures and apply them to solve one ILP problem as well as the RAP in Equation (8.11). The B&B algorithm contains the procedure of implicit enumeration. Even for the 0-1 LP problem, the number of possible solutions is 2^n, where n is the number of decision variables. The main idea of implicit enumeration is to skip the enumeration of a large part of the solutions. The underlying mechanism used in implicit enumeration is 'divide and conquer'. There are two major phases in B&B: In the phase of separation (i.e. branching), the solution set is divided into subsets; in the phase of evaluation (bounding), the subset is evaluated using LP techniques and those subsets that do not contain the optimal solution are eliminated. An ILP is an LP with additional integrity constraints, and the optimal value of the LP is an upper bound or lower bound of the ILP for a maximization or minimization problem, respectively.

Prior to the B&B algorithm, we first present the search tree, i.e. the B&B tree, which is useful for the illustration of the algorithm procedures. For simplicity, let all decision variables be binary numbers $\boldsymbol{x}\in\{0,1\}^n$. As shown in Figure 8.2, at the top node of the tree, we set $x_1=0$ for its left branch and $x_1=1$ for its right branch. For each branch, we solve the relaxation of the original ILP problem with $x_1=0$ and $x_1=1$, respectively. Then we proceed to the second variable x_2 under the branches generated by x_1, and repeat the branching and evaluation steps for x_2. This process iterates until the last variable is explored. Finally, a search tree is generated, as shown in Figure 8.2.

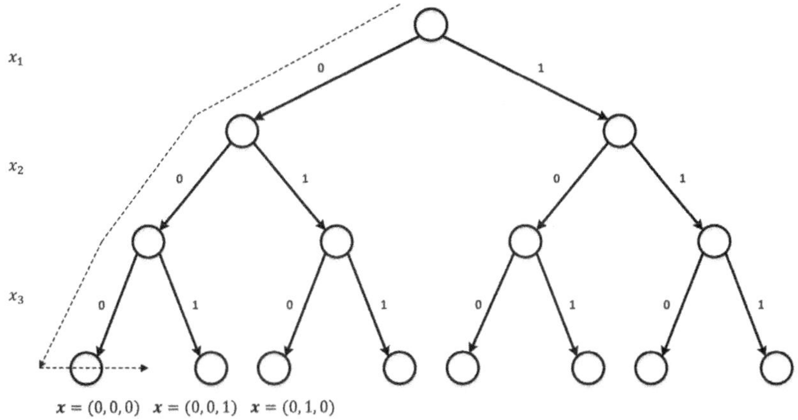

Figure 8.2 Search tree.

Besides the branching and evaluation, the B&B algorithm also involves bound and cut procedures. The whole algorithm is presented as follows:

Step 1. **Initialization**: solve the relaxation of the original problem.

Step 1.1. If all elements of the solution vector are integers, then stop; and
Step 1.2. If any element of the solution vector is non-integer, then store this solution as one bound to the original problem.

Step 2. **Branch**: Select one appropriate variable x_i, set its value to the closest upper and lower integers and obtain two sub-problems, respectively.
Step 3. **Bound**: Select and solve one relaxed sub-problem.
Step 4. If any of the following condition is satisfied:
- *the solution is an integer solution;*
- *the corresponding problem is infeasible;*
- *the solution is worse than a known feasible solution;*

then stop the branch and then go to Step 3;
Otherwise, go to Step 2.
Step 5. Stop the algorithm till all nodes (sub-problems) in the search tree are visited.

Example 8.3 Consider the following Knapsack problem:

$$\min_{x_1,x_2,x_3,x_4} \quad 9x_1 + 2x_2 + 4x_3 + 3x_4 \tag{8.13a}$$

$$\text{s.t. } 3x_1 + 2x_2 + 4x_3 + 2x_4 \leq 8 \tag{8.13b}$$

$$x_i \in \{0,1\}, \ i = 1,2,3,4 \tag{8.13c}$$

The B&B solution procedures are illustrated in Figure 8.3. The order of branching is x_1, x_2, x_3, x_4, and the relaxation of the original problem is as follows:

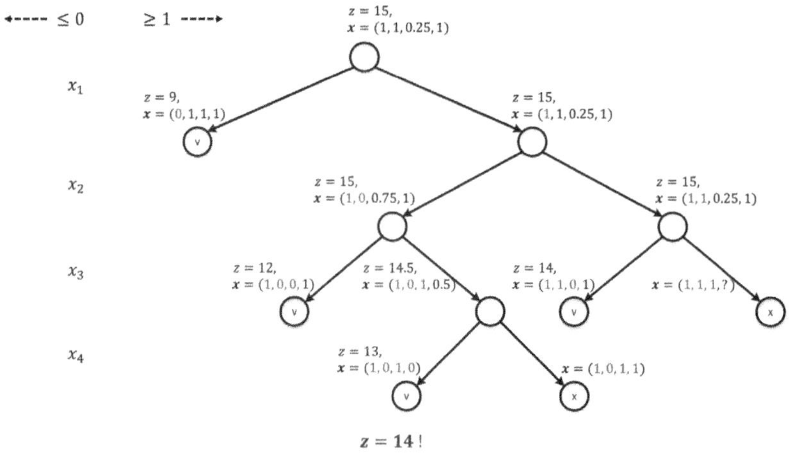

Figure 8.3 The search tree of Example 8.3.

$$\min_{x_1,x_2,x_3,x_4} \; 9x_1 + 2x_2 + 4x_3 + 3x_4 \tag{8.14a}$$

$$\text{s.t.}\, 3x_1 + 2x_2 + 4x_3 + 2x_4 \le 8 \tag{8.14b}$$

$$0 \le x_i \le 1,\; i = 1,2,3,4 \tag{8.14c}$$

At the first step, we solve the problem in Equation (8.14) and find the optimal solution $x = (1,1,0.25,1)$ to which corresponds the objective value $z = 15$. Then we generate the branches at x_1. By setting $x_1 = 0$, we obtain the sub-problem (8.15a)-(8.15c) for the left branch, and its optimal solution is $x = (1,1,1)$, which corresponds to the objective value $z = 9$. This is a feasible and integer solution, and we record it as the temporary optimal solution. We can stop the branch from this node because the solutions to all sub-problems under this node will not be better than the present optimal solution.

$$\min_{x_2,x_3,x_4} \; 2x_2 + 4x_3 + 3x_4 \tag{8.15a}$$

$$\text{s.t.}\, 2x_2 + 4x_3 + 2x_4 \le 8 \tag{8.15b}$$

$$0 \le x_i \le 1,\; i = 2,3,4 \tag{8.15c}$$

Then we look at the sub-problem of the right branch of x_1. By setting $x_1 = 1$, we will get the same solution as the one to the original problem. Then we continue the branching and evaluation process following the depth-first rule [8]. For the sub-problem whose $x_1 = 1, x_2 = 0$ and $x_3 = 0$, we can obtain $x_4 = 1$ as the optimal solution, which corresponds to the objective value $z = 12$. Because it is larger than the objective value of the temporary optimal solution, we take this solution as the newest optimal solution. Also, we stop the branching at this node.

The B&B process continues till all nodes of the search tree are visited. The global optimal solution is $\boldsymbol{x} = (1,1,0,1)$, which corresponds to the objective value $z = 14$. The sub-problem whose solution is $x_1 = 1, x_2 = 1$ and $x_3 = 1$ becomes infeasible because the constraint will require x_4 being negative. Thus, the branching process stops on this node.

Next, we apply the B&B method in reliability optimization. Reliability optimization problems are typically non-linear. On the other hand, they are usually monotone to the decision variables, e.g. the objective function value of RAP in Equation (8.11) increases as x_{ij} increases. Thus, their relaxations would be convex in several cases. If the relaxation is convex the original non-linear IP problem can be solved by the B&B method. Under this situation, we have two extensions to Example 8.3: The decision variables non-binary integers, and the objective function is non-linear. The key steps of the B&B method to solve this problem are presented as follows:

1) Choose one variable that has a non-integer value and branch the variable to the next higher integer for one sub-problem and the next lower integer for the another sub-problem. The real value of variable j can be expressed as $x_i = \lfloor x_i \rfloor + x^*$, where $\lfloor x_i \rfloor$ is the integer part of x_i and $0 < x^* < 1$. The lower-bound and upper-bound constraints of the two mutually exclusive problems are $x_i \geq \lfloor x_i \rfloor + 1$ and $x_i \leq \lfloor x_i \rfloor$, respectively. Add these two constraints to both branched problems. Solve both problems by non-linear optimization methods, e.g. KKT conditions.
2) Now x_i is an integer in either branch. Fix the integers of x_i for the following steps of B&B. Select the branch that results in the highest system reliability. Then repeat the above steps on another variable $x_k \neq x_i$ for each new problem until all variables become integers.
3) Stop branching the problem if the solution is worse than the current best integer solution. Stop the iteration when all the desired integer variables are obtained.

The following example illustrates these procedures.

Example 8.4 Given the parameters in the following table, solve the RAP in Equation (8.11). This is a simplified problem where each subsystem has only one component type.

Subsystem, i	**1**	**2**	**3**	**4**
r_i	0.8	0.7	0.75	0.85
c_i	1.2	2.3	3.4	4.5
$C_0 = 56$				

$$\max_{x_1, x_2, x_3, x_4} R(x) = \left(1 - 0.2^{x_1}\right)\left(1 - 0.3^{x_2}\right)\left(1 - 0.25^{x_3}\right)\left(1 - 0.15^{x_4}\right) \tag{8.16a}$$

$$\text{s.t.} 1.2x_1 + 2.3x_2 + 3.4x_3 + 4.5x_4 \leq 56 \tag{8.16b}$$

$$1 \leq x_i, x_I \in Z^* \tag{8.16c}$$

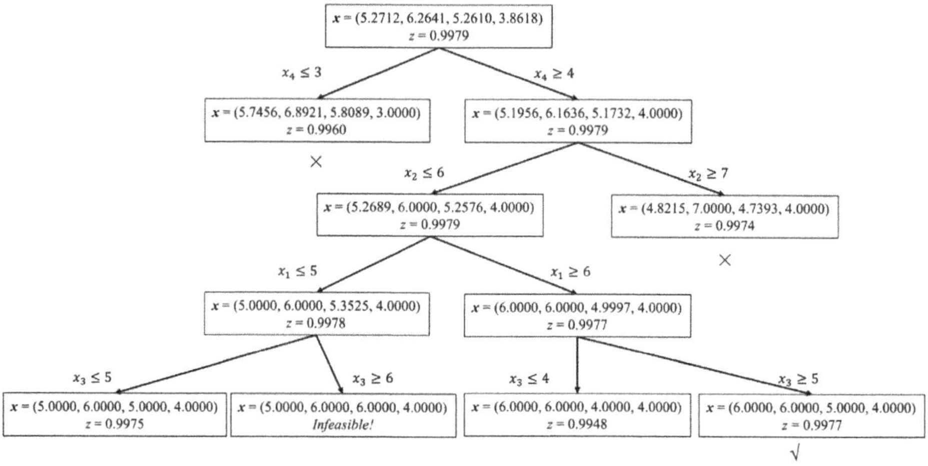

Figure 8.4 The search tree of Example 8.4.

Because the relaxation of the objective function is convex, the B&B method is applicable to this problem. The solution procedures are illustrated in the search tree in Figure 8.4. In Step 1, the original problem is relaxed and solved. The optimal solution found is $x = (5.2712, 6.2641, 5.2610, 3.8618)$, which corresponds to the objective value $z = 0.9979$. Continuous convex optimization problems can be solved by various computation software packages, e.g. MATLAB. For details about convex optimization, the interested readers can refer to the classical book by Boyd and Vandenberghe [1].

In Step 2, we choose x_4 for branching. Because the optimal x_4 is 3.8618 in the last step, we generate two branches: the left for $x_4 \leq 3$ and the right for $x_4 \geq 4$. For the two sub-problems under these branches, we set $x_4 = 3$ and $x_4 = 4$, respectively.

In Step 3, we relax both sub-problems under the two branches, solve the relaxed problems and obtain the optimal solutions $x = (5.1956, 6.1636, 5.1732, 4.0000)$ with $z = 0.9979$, and $x = (5.7456, 6.8921, 5.8089, 3.0000)$ with $z = 0.9960$, for the right and left sub-problems, respectively. Because the right sub-problem has a higher optimal value, we give priority to this sub-problem for further branching. This is a 'greedy search'-type of heuristic that follows the large-first principle for maximization.

The process continues until we obtain a feasible optimal solution $x = (6.0000, 6.0000, 5.0000, 4.0000)$ with $z = 0.9977$. Because this optimal value is higher than those of the sub-problems under the branch $(x_4 \leq 3)$ and the branch $(x_4 \leq 3,\ x_2 \geq 7)$, we stop further branching at the respective nodes.

At last, the optimal redundancy allocation scheme is $(6, 6, 5, 4)$. This is the true global optimal solution to this example.

8.3 Exercises

1) Solve Example 8.2 by the graphical method.

Solve the following RAP given the parameters in the table below:

Stage, j	1	2	3	4
r_j	0.8	0.7	0.75	0.85
c_{1j}	1.2	2.3	3.4	4.5
c_{2j}	5	4	8	7
$b_1 = 56$				
$b_2 = 120$				

$$\max R_s(\mathbf{x}) = \prod_{j=1}^{4}\left[1-(1-r_j)^{x_j}\right]$$

$$\text{s.t.} \sum_{j=1}^{4} c_{ij}x_j \le b_i, i = 1,2$$

$$1 \le x_j, j = 1,2,3,4$$

In the B&B method, the order of branching is important to the computation complexity. Change the branching sequence in Example 8.3 to '$x_3 \to x_4 \to x_2 \to x_1$' and resolve again the problem. Is this sequence more efficient? Why?

References

1 Boyd, S. and Vandenberghe, L. (2004). *Convex Optimization*. Cambridge, UK: Cambridge University Press.

2 Bertsimas, D. and Weismantel, R. (2005). *Optimization over Integers*. Belmont, MA: Dynamic Ideas.

3 Wolsey, L.A. (1998). *Integer Programming*. New York City: John Wiley & Sons.

4 Kulshrestha, D. and Gupta, M. (1973). Use of dynamic programming for reliability engineers. *Reliability, IEEE Transactions On* 22 (4): 240–241.

5 Dantzig, G.B. (1963). *Linear Programming and Extensions*. Princeton, NJ: Princeton University Press.

6 Ghare, P. and Taylor, R. (1969). Optimal redundancy for reliability in series systems. *Operations Research* 17 (5): 838–847.

7 Beraha, D. and Misra, K. (1974). Reliability optimization through random search algorithm. *Microelectronics Reliability* 13 (4): 295–297.

8 Tarjan, R. (1972). Depth-first search and linear graph algorithms. *SIAM Journal on Computing* 1 (2): 146–160.

9

Evolutionary Algorithms (EAs)

Practical reliability optimization problems can be, in general, difficult to solve via standard mathematical programming methods, e.g. linear programming (LP) and integer programming (IP) because the objective and constraints are mostly non-linear, and the decision variables can mix continuous variables (e.g. test interval) and integer variables (e.g. redundancy). Evolutionary algorithms (EAs), inspired by natural principles of evolution, can perform population-based stochastic search to produce good solutions (not necessarily globally optimal) in polynomial time. In addition, EAs are easy to comprehend and, thus, to implement. As a result, many reliability engineers and researchers often resort to EAs for solving their reliability optimization problems.

In literature, EAs are also called meta-heuristics. These two terminologies are used interchangeably in many occasions, so as in this book. In practice, mathematical programming methods might be able to find the global optimal solution of a complex reliability optimization problem, but the computation cost can be high and the achieved global optimal solution can be slightly better than other local optimal solutions. Under these situations, EAs can be the preferred alternatives. EAs generally contain two components: randomization and improvements on the best solutions. Randomization aims at avoiding the solutions being trapped in local optima and increasing diversity. The improvements on the best solutions control the direction of random search so that the solutions converge to optimality. The balance between randomization and selection of the best solutions is the key to the success of EAs in application.

Generally, EAs possess the following four advantages:

1) They are readily working with continuous, integer, categorical and mixed decision variables.
2) They search from a population of solutions, instead of a single one.
3) They require only evaluating the objective function (without calculating its deviation or derivative).
4) They use probabilistic transition rules to guide the search, which helps avoid getting trapped in local minima.

For more advanced knowledge about EAs, please refer to the books [1,2].

Reliability Analysis, Safety Assessment and Optimization: Methods and Applications in Energy Systems and Other Applications, First Edition. Enrico Zio and Yan-Fu Li.

9.1 Evolutionary Search

Evolutionary search is the foundation of EAs. It combines Monte Carlo random search and the evolution strategy that iteratively improves the obtained solution. The Monte Carlo random search procedure is straightforward: It generates a large but finite number of random samples of the decision variables in a way that they are uniformly distributed in the domain of interest. After evaluating the corresponding objective function of all generated samples, the Monte Carlo search returns the best solution found randomly in the domain of interest.

Let us consider the following 0-1 Knapsack problem:

$$\underset{x_1,x_2,\ldots,x_m}{\text{Max}} \; f(x)=\sum_{i=1}^{n} w_i x_i \tag{9.1a}$$

$$\text{s.t.}\; g(x)=\sum_{i=1}^{n} c_i x_i \leq C_0 \tag{9.1b}$$

$$x_i \in \{0,1\} \tag{9.1c}$$

where w_i, c_i and C_0 are all constants. This is an NP-hard problem, e.g. if $n=500$, and its search space is of size $2^{500} \gg 10^{80}$, i.e. the number of elementary particles in the universe. Let $\boldsymbol{x}=(x_1,x_2,\ldots,x_n)$ denote the vector of candidate solution, $\boldsymbol{x}^*$ denote the global optimal solution, and K_{max} denote the maximal number of iterations of the search. The Monte Carlo search procedures for solving this problem are presented as follows

---

Algorithm 9.1 Monte Carlo search

1. Set $\boldsymbol{k} \leftarrow 0$, $\boldsymbol{x} \leftarrow (0,0,\ldots,0)$
2. For ($\boldsymbol{t} \leftarrow \boldsymbol{t}+1$, $\boldsymbol{t} < \boldsymbol{T}_{\boldsymbol{max}}$)
3. Randomly sample a solution $\boldsymbol{x} \in \{0,1\}^{\boldsymbol{n}}$
4. If $\boldsymbol{x}=\boldsymbol{x}^*$ then
5. Terminate and return $\boldsymbol{x}$
6. End-if
7. End-for

The disadvantage of the Monte Carlo random search mainly lies in the fact that in each iteration, one candidate solution is generated independently and randomly for trial, without utilizing the current best solution. This type of strategy is intuitively unpromising because it could even generate repeated solutions. In the following paragraph, a quantitative analysis on the computational complexity of this method is provided.

In the Monte Carlo search, the probability to generate $\boldsymbol{x}^*$ at a single iteration is $P_1=\Pr(\boldsymbol{x}=\boldsymbol{x}^*)=2^{-n}$. The probability to generate $\boldsymbol{x}^*$ with k iterations is $P_k=1-(1-2^{-n})^k$. Solving for k, the number of iterations needed to find $\boldsymbol{x}^*$ is

$k = \ln(1-P_k)/\ln(1-2^{-n})$. Given the approximation $x \approx \ln(1+x)$, we have $k \approx -2^n \ln(1-P_k)$. It is obvious that k is exponential to n, which means the Monte Carlo search can become an extensive method from the computational viewpoint. The condition that the Monte Carlo method uses more iterations for finding the optimal solution than the enumeration does is $-2^n \ln(1-P_k) \geq 2^n$, which says that $P_k \geq 1-e^{-1} \approx 0.63$. This result means that if we want to have a probability higher than about 0.63 to find the optimal solution in k runs, then enumeration is better than Monte Carlo search.

Evolutionary search, on the other hand, takes the advantage of randomization and creates new solutions based on the present best solution. The following (1+1) EA [3] is among the simplest EAs. The shaded area is the main difference between this algorithm and Monte Carlo search, and the random modification is carried out on the best solution of the last iteration and the present best solution is passed to the next iteration.

Algorithm 9.2 (1+1) Evolutionary Algorithm

```
1.  Set t ← 0, x ← (0,0,...,0)
2.  Randomly sample a solution x ∈ {0,1}^n
3.  For (t ← t+1, t < T_max)
4.    Create a copy x' of x
5.    Invert each bit of x' with probability p
6.    If (x' matches x* in more bits than x) then
7.      x ← x'
8.    End-if
9.    If x = x* then
10.     Terminate and return x
11.   End-if
12. End-for
```

To analyze the computational time of (1+1) EA for finding the global optimal solution x^*, assume that m bits are still not optimal in the best solution. The probability to preserve all $n-m$ correct bits under Step 5 is $(1-p)^{n-m}$. The probability to improve exactly one of the wrong bits is $mp(1-p)^{m-1}$. The probability that the new solution x' is better than the previous one x is $\Pr(x' \text{ is better than } x) \geq mp(1-p)^{n-1}$. The expected number of iterations until an improvement happens is $E_1 \leq \frac{1}{mp(1-p)^{n-1}}$. Thus, the expected total number of improvements is $E_n = \sum_{i=1}^{n} E_i \leq \frac{1}{p(1-p)^{n-1}} \sum_{m=1}^{n} \frac{1}{m}$. Consider that $\lim_{n\to\infty}\left(\sum_{m=1}^{n}\frac{1}{m} - \ln n\right) = \gamma = 0.0522\ldots$, we have $E_n \approx \frac{1}{p(1-p)^{n-1}}\ln(n)$. Assume that $p = \frac{1}{n}$, finally we get $E_n \approx ne\ln(n)$. In conclusion, (1+1) EA is of $n\log(n)$ computational complexity. In this algorithm, there are only one-bit improving mutations and an upper bound on E_n. For more advanced EAs, e.g. genetic algorithm, the expected computational complexity would be much smaller.

To summarize, there are two basics components in EA: randomization and selection of the best solutions. The former avoids the solutions being trapped in local optima and increases diversity, and the latter ensures the solutions converge to optimality. Balance between the two components is the key to the success of EA implementations.

9.2 Genetic Algorithm (GA)

A genetic algorithm (GA) [4] is perhaps the most popular and successful EA. It imitates the biological evolution and natural selection processes on a group of individuals (solutions), to eventually achieve quality solutions. The standard operation procedures of a single-objective GA (SOGA) are presented as follows:

Algorithm 9.3 Genetic Algorithm

1. Set $\boldsymbol{t} \leftarrow 0$
2. Initialize $\boldsymbol{X}_t$
3. Evaluate $\boldsymbol{X}_t$
4. For $(\boldsymbol{t} \leftarrow \boldsymbol{t}+1, \boldsymbol{t} < \boldsymbol{T}_{max})$
5. Select $\boldsymbol{X}_t$ from $\boldsymbol{X}_{t-1}$
6. Crossover $\boldsymbol{X}_t$
7. Mutate $\boldsymbol{X}_t$
8. Evaluate $\boldsymbol{X}_t$
9. Apply elitist strategy to $\boldsymbol{X}_t$ (given $\boldsymbol{X}_{t-1}$)
10. End-for

where $\boldsymbol{X}_t$ is the population of solutions (i.e., individuals), t is the generation counter (i.e., the population iteration index), and G_{max} is the maximum number of generations. $\boldsymbol{X}_0$ consists of a group of encoded individuals $\boldsymbol{x}$ randomly generated at the initialization step. The evaluation of the population requires computing the value of the objective function for each individual solution in the population and converting it into the fitness value, which reflects the quality of the corresponding individual. The selection step determines the group of individuals entering the evolution process with a probability related to the fitness values. The individuals with high fitness values have large probabilities to survive in the evolution process. The crossover and mutation are the important evolutionary operators of GA. In the crossover, generally, two individuals, named parents, are paired to produce new individuals, named offsprings, by exchanging some parts of the encoded solutions. Crossover allows the parts of the good solutions to be retained and copied in the population, so the algorithm can eventually converge to an overall good solution. On the other hand, the mutation operator randomly changes the coding of the individuals. The perturbations brought by mutation introduce diversity into the population and assist the search escaping from local minima. The elitist strategy maintains the fittest individual of the population. It ensures that the best solution will not be lost during the stochastic search process. The following sections introduce each main step of GA.

9.2.1 Encoding and Initialization

To design a GA, the first step is to translate the candidate solutions into *genotypes* for manipulation by GA. There are different ways to carry out this task. The method chosen must be relevant to the problem that is being solved. The encoding will influence the fitness evaluation and the genetic operators. In general, there are four types of encoding: discrete encoding, real-valued encoding, order-based encoding, and tree-based encoding, each suitable to a certain problem type.

Discrete encoding makes use of discrete values, e.g. binary, integer, or any other system with a finite set of values. The most common discrete encoding is the binary encoding. Figure 9.1 shows one example of binary chromosome and its various phenotypes. In analogy to the chromosome and gene of living creatures, GA also considers chromosomes that represent candidate solutions and genes that are the elements of the chromosomes. A chromosome is a genotype. To evaluate the fitness of the chromosome, it needs to be translated into a phenotype, which is the original form of the candidate solution. In Figure 9.1, three different phenotypes are shown, corresponding to integer, real number and assignment type of solutions, respectively. The solution type is determined by the nature of the optimization problem. Figure 9.1 also implies the potential difficulty of binary encoding in dealing with continuous search domains of large dimensions and high numerical precision. One problem can occur when a variable takes a finite number of values, which is not a power of 2, because in this case some binary chromosomes are redundant or useless.

The real-valued encoding is a straightforward type of encoding. Generally, it directly utilizes the original form of a candidate solution, i.e. a vector of the decision variables (real-valued), $\boldsymbol{x} = (x_1, \ldots, x_n)$ where $x_i \in R$. Thus, real-valued encoding appears to be a natural alternative for solving optimization problems with decision variables in continuous domains.

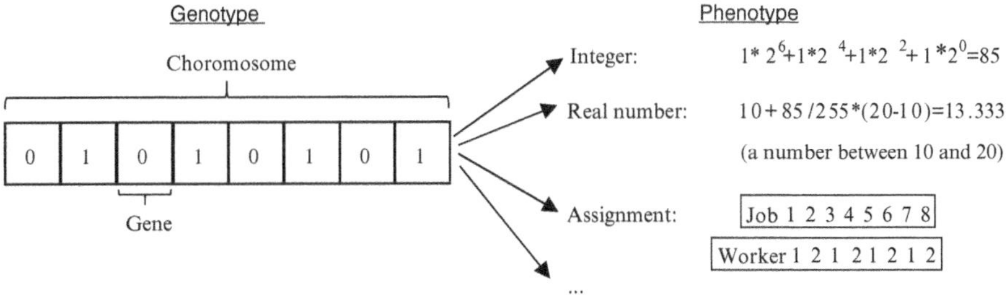

Figure 9.1 Example of binary encoding.

In the order-based encoding scheme, individuals are represented as permutations used for ordering/sequencing problems. For this type of encoding, special genetic operators are needed to make sure the individuals remain valid permutations. For example, to solve a travelling salesman problem, each city can be assigned a unique number from 1 to 5, and thus, a candidate solution could be (5, 4, 2, 1, 3).

The tree-based encoding is mainly used for solving the optimization problems that can be formulated in terms of finding an optimal tree structure. For example, for the shortest path problem with multiple sources and destinations, in the encoding scheme, the individuals in the population are trees. Each tree represents one path from a source node to several destination nodes, i.e. the source is the root and the destinations are mapped into leaf nodes. Figure 9.2 shows such an example and the related tree, representing six paths from the source node A to the destination nodes D, E, F, and G.

Initialization is performed on the encoding scheme. It generates the initial population of chromosomes (i.e. solutions) uniformly distributed over the search space. For example, for binary encoding, the initialization samples the value 0 or 1 with probability 0.5 for each gene; for real-valued encoding, the initialization uniformly samples a real value within a given interval for each gene. The above procedures create a population from scratch. In other cases, the initial population can be inherited from previous results or other heuristics; in these cases, the initial population is closer to the optimal solution but could lead to possible loss of genetic diversity and introduce possible unrecoverable bias in the search.

9.2.2 Evaluation

Evaluation is the step in which the fitness value of a chromosome is calculated. The fitness value drives the probability that a chromosome survives in the selection procedure to the next generation. To compute the fitness value, the evaluation step first

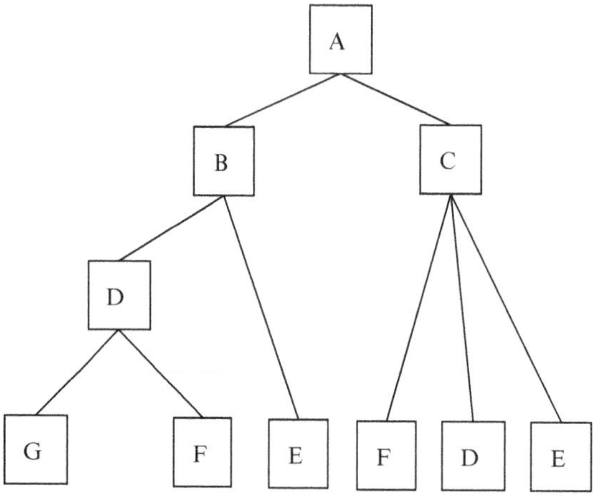

Figure 9.2 Example of tree-based encoding.

decodes the chromosome, i.e. translates the genotype into the phenotype, and then uses the decoded value to compute the value of the objective function. In general, if the objective function is to be maximized, then no further conversion of the value is needed; if the objective function is to be minimized, then the objective value needs to be converted, e.g. via taking the reciprocal $1/f(x)$, so that in the selection procedure, the chromosomes with larger fitness values have higher survival probabilities to the next generation.

In addition, the constraints of the optimization problem need to be considered in the evaluation step because it could happen that one phenotype breaks some constraints. Typically, there are two ways to handle constraints: penalizing the fitness and implementing a repair method. The former simply adds to the fitness value a term that measures the degree of the constraint violation, e.g. $\delta \max(g(x) - C_0, 0)$ in Equation (9.1) where δ is the penalty coefficient, a large constant value. The latter involves designing a mechanism that can fix the chromosome to satisfy all constraints, e.g. via switching some '1' valued genes to '0', in Problem 9.1.

As a matter of fact, the evaluation step is generally the most computationally expensive step for real applications because the objective value might be the computed outcome of a subroutine, a black-box simulator, or of any external process (e.g. robot experiment). Thus, it is recommended to avoid re-evaluating the same chromosome throughout all generations. Another option is to use computationally cheaper surrogates to approximate the fitness evaluations, but this could disturb the evolution process towards the optimal solution, the surrogates cannot be used for a large number of generations, and the evolution path by using surrogates needs to be constantly checked and corrected if needed.

9.2.3 Selection

After the fitness evaluation, the selection strategy is applied to favor the best chromosomes (i.e., with the highest fitness values) to have more chance of surviving to the following genetic operations. Different from the genetic operations, which aim at creating diversified offsprings, selection represents the evolution pressure that forces the population to evolve toward an improved one. Still, the chromosomes of lower quality must also have a certain chance of survival to the next generation because they may carry useful genetic elements.

Fitness-proportional selection is the most popular method for selection. It is also referred to as roulette-wheel selection. For one chromosome, the probability of being selected is determined as

$$p_i^s = \frac{f_i}{\sum_i f_i}. \qquad (9.2)$$

Under this scheme, a chromosome of higher fitness value has more chances to be selected.

However, this method has three disadvantages: 1) the risk of trapping into local optima because the superior (fittest) chromosomes could quickly dominate the entire population; 2) the discrimination power is low when the fitness values are close to each other; and 3) the proportion reflecting the relative share of each chromosome in the minimization problem should be distorted due to the inversion of the fitness values.

To remedy these drawbacks, other selection approaches, e.g. tournament selection and rank-based selection, can be used. Tournament selection [5] performs several tournaments among a few chromosomes randomly chosen from the population and the winner of each tournament is selected to undergo genetic operations. Compared to the fitness-proportional selection procedure, tournament selection is independent from the scale of the fitness values. Rank-based selection [6] arranges the chromosomes from their best to the worst fitness values, and selects the chromosomes according to a probabilistic function of their ranks where the best chromosome is ranked m (total number of chromosomes in the population) and the worst chromosome is ranked 1. This basically overcomes the three major drawbacks of the fitness-proportional selection.

9.2.4 Mutation

The mutation is a genetic operator that alters one or more gene values in a chromosome so the genetic diversity is achieved. The mutation is made according to a user-defined mutation probability p_m (typically a low value, e.g. $p_m \leq 0.001$). An example is the inversion operation of one bit in the (1+1) evolutionary algorithm in Section 9.1. A mutation can significantly change the previous solution. For different encoding schemes, the mutation operators are different.

For binary encoding, the mutation operator inverts the value of a gene, with a given probability p_m. This is identical to the inversion operation in the (1+1) EA.

For real-valued encoding, the mutation operator perturbs values by adding a random noise to the original gene values. Typically, a normally distributed noise $N(0,\sigma)$ is used where 0 is the mean value and σ is the standard deviation. Thus, the mathematical expression for mutation of the i-th gene is $x_i^{'} = x_i + N(0,\sigma_i)$. Another type of mutation operator is the uniform mutation, which replaces the value of the chosen gene with a uniform random value selected within user-defined bounds for that gene. Both these mutation operators can be used for real-valued and integer-valued encoding schemes.

For order-based encoding, a standard mutation operator leads to infeasible solutions. Therefore, at least two values must be changed at the same time. Under this setting, p_m now represents the probability the operator will be applied once on the whole chromosome rather than individually on each gene. The swap operator, shown in Figure 9.3, randomly selects two genes and then swaps their positions. More information about other order-based mutations, e.g. insert mutation and inversion mutation can be found in Eiben and Smith's book [7].

For tree-based encoding, the mutation operator randomly selects a node in the tree structure and replaces its subtree (node inclusive) with a randomly created tree.

Figure 9.3 Swap operator for order-based encoding.

Similar to the order-based case, p_m represents the probability that a whole chromosome is selected for mutation. A schematic example of this mutation is shown in Figure 9.4. This type of mutation operator is also called subtree mutation. For details about other tree-based mutations, please refer to Koza's genetic programming book [8].

To conclude the mutation section, know that a mutation operator should allow reaching each part of the search space, the amount of mutation is important and should be controllable, mutation should produce physically valid chromosomes.

9.2.5 Crossover

Crossover, also called recombination, is a genetic operator that combines the genetic information of the paired parent chromosomes to generate new offsprings (i.e. children) as an analogy to the biological reproduction. Similar to mutation, crossover occurs according to a user-defined probability p_c (typically a high value, e.g. $p_c \geq 0.6$), which determines the chance of applying the crossover operator on each pair. For different encoding schemes, the applicable crossover operators are different.

For binary-valued, integer-valued, and real-valued encodings, the same type of crossover operators can be applied. Figure 9.5 shows the single-point crossover scheme where the paired chromosomes are cut at the same crossover point, and the genes to the right

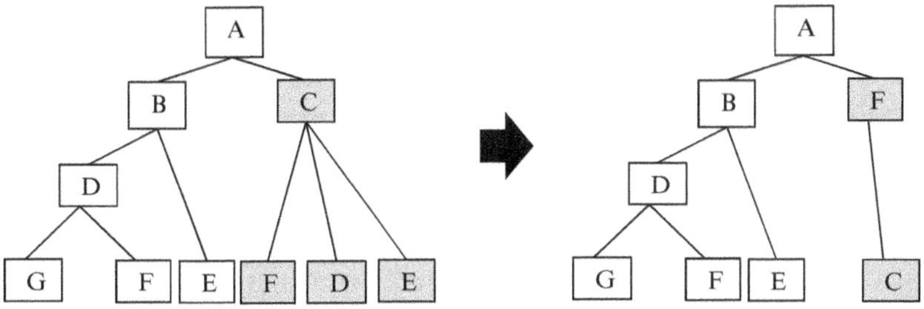

Figure 9.4 One example of tree-based mutation.

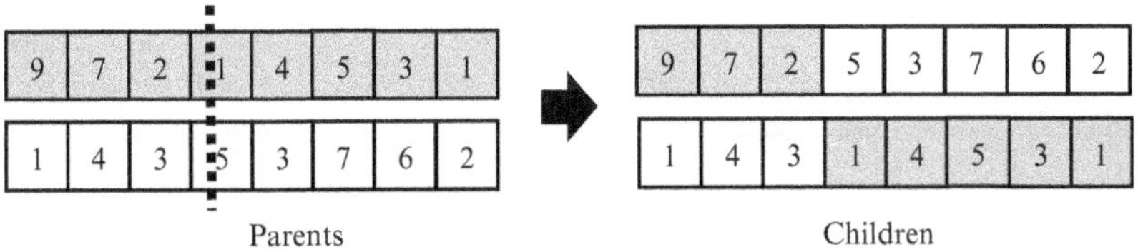

Figure 9.5 Single-point crossover.

of this point are swapped between the two parent chromosomes. The crossover point occurs with an equal probability between any two adjacent genes.

Two-point crossover is the extension of single-point crossover. The two crossover points are randomly selected and the genes between these two points are swapped between the parent chromosomes. One example is shown in Figure 9.6. Furthermore, two-point crossover can be generalized to n-point crossover, which swaps the genes between two adjacent crossover points between the parents.

Uniform crossover operator swaps the genes on the randomly selected crossover points between the parents. To illustrate this crossover operator, as shown in Figure 9.7, a mask vector of '0' and '1' values is randomly generated, and the genes of the parents are swapped at the points with '1' values on the mask vector. There exists a relation between the uniform crossover and n-point crossover. For example, a mask '00011100' defines a two-point crossover with the fourth and sixth genes as crossover points.

Because crossover is conducted with high probability, its *disruptive effect* needs to be considered carefully when selecting or designing the crossover operators for a GA. For example, in the global optimal solution, there are some inseparable building blocks, e.g. composed of two consecutive genes. The single-point crossover (or mutation) operator is said to disrupt the building block if it separates the two genes and distrutes them into different children and then the building block might not appear in neither child. This effect can be remedied by utilizing the two-point crossover, that does not break the block. However, the disruptive effect is not always unwelcomed because it can increase the chance for the GA to jump to different points of the search space, which enhances the exploration capability. In the end, as said before, the right balance between global exploration and local exploitation is the key to the successful implementation of a GA. For details about multi-point crossover (including n-point crossover and uniform crossover) and disruptive effects, the interested readers can refer to De Jong and Spear's work in [9].

For real-valued encoding, a few special crossover operators are different from the common single-point, multi-point, and uniform crossover operators. One example is the arithmetical crossover where each gene of the offspring chromosome is the weighted sum of the parent genes at the same position. Mathematically speaking, let $\boldsymbol{x}_1$ and $\boldsymbol{x}_2$ denote the two parent chromosomes and $\boldsymbol{x}'_1$ and $\boldsymbol{x}'_2$ denote the two children chromosomes after arithmetical crossover: Then, for the i-th gene x'_{1i} in the first child, we have $x'_{1i}=\lambda x_{1i}+(1-\lambda)x_{2i}$; similarly for the i-th gene x'_{2i} in the second child, we have $x'_{2i}=\lambda x_{2i}+(1-\lambda)x_{1i}$ where λ is a constant for uniform arithmetical crossover and varies with the generations for non-uniform arithmetical crossover. For other types of real-valued crossover operators, the interested readers can refer to [10].

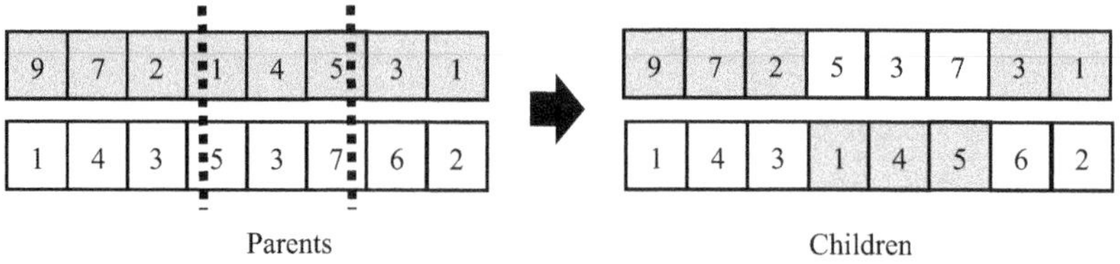

Figure 9.6 Two-point crossover.

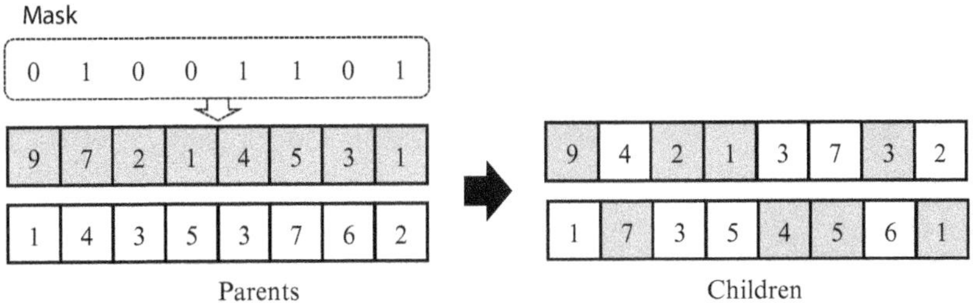

Figure 9.7 Uniform crossover.

For order-based encoding, similar to the case of mutation, standard crossover operators, e.g. single-point and n-point operators, may lead to infeasible solutions. Thus, various specialized operators are designed for order-based encoding. Among them, order 1 crossover is a common one; its key idea is to preserve the relative order of the occurrence of the genes.The procedure is presented as follows:

Procedure 9.1 Order 1 crossover

1. Randomly choose a gene set from the first parent and copy this to the first child.
2. Copy the remaining genes, that are not in the copied part, to the first child:
   2.1 Starting right from the cut point of the copied part,
   2.2 Using the order of the same genes in the second parent, and
   2.3 Wrapping around at the end of the chromosome.
3. Repeat this process with the parent roles reversed

Figure 9.8 Illustrative example of the procedure.

More information about other crossover operators, e.g. partially mapped crossover and cycle crossover can be found in Eiben and Smith's book [7].

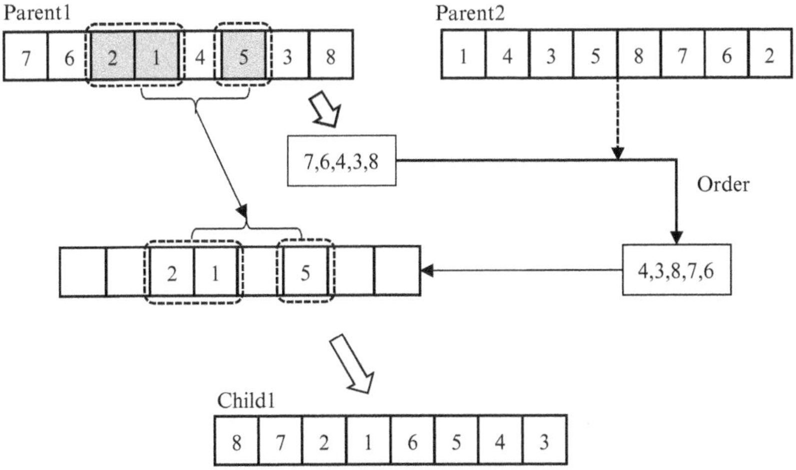

Figure 9.8 Example of Order 1 crossover.

For tree-based encoding, the crossover operator randomly selects one node in each parent tree and swaps the subtrees between the parents under the respective nodes. A schematic example of this crossover is shown in Figure 9.9. This type of crossover operator is also called subtree crossover. For details about other tree-based crossover operators, please refer to Koza's genetic programming book [8].

To conclude the crossover section, know that the offspring should inherit certain genetic materials from the parents, the crossover operators should be designed in conjunction with the representation scheme, and crossover should produce valid chromosomes.

9.2.6 Elitism

Elitism, i.e. elitist selection, allows a limited number of the chromosomes with the best fitness values in the present population to be preserved in the next generation. Typically, the elitism rate is used to control the proportion of the selected best chromosomes from the current population. For example, given a population of 20 chromosomes, elitism rate = 0.1 means the top two chromosomes will survive unaltered in the next generation.

Elitism is a simple but non-negligible step in GA, and furthermore, it has been proven in theory that without elitism, GA cannot guarantee to converge to the global optimal solution [11].

9.2.7 Termination Condition and Convergence

Because GA is a stochastic optimization method, the termination condition needs to be decided before its implementation. The maximal number of generations (or fitness

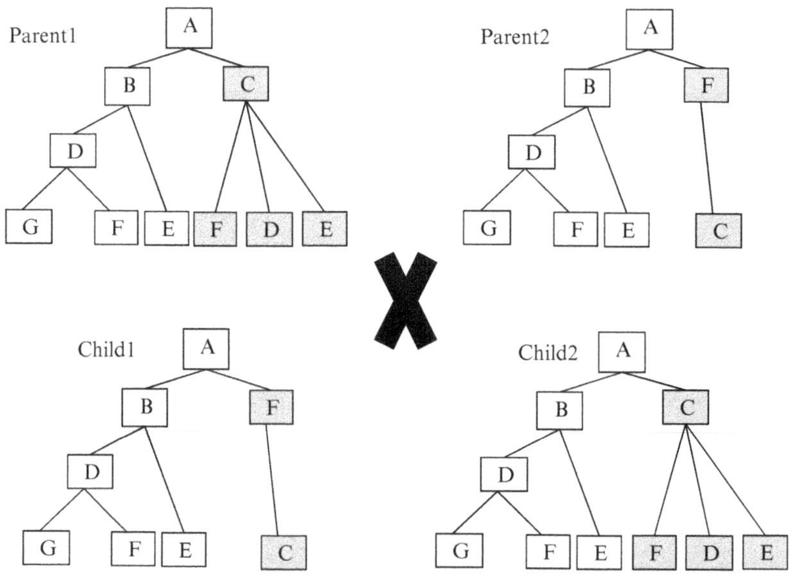

Figure 9.9 Example of crossover for tree-based encoding.

evaluations) is one commonly used condition. It is mainly related to the limited computational resources. Another typical type of termination condition is the maximal number of generations with unchanged best fitness value. Let f_t^* and f_{t-1}^* denote the best fitness values from the population at generation t and $t-1$, respectively, and let $\delta_t = f_t^* - f_{t-1}^*$ be the difference between two generations: if $\delta_t = 0$ for a predefined number T_{conv} of generations, then the algorithm will terminate. Moreover, the above two conditions can be used simultaneously to achieve the balance between the computational resource and the convergence requirement.

9.3 Other Popular EAs

This section briefly introduces two EAs: differential evolution (DE) and particle swarm optimization (PSO), which are as successful as GAs in solving various problems. DE was originally proposed as a population-based global optimization algorithm for real-valued problems [12]. Due to its simplicity and efficiency, DE has soon become a popular EA in various scientific and engineering fields, including reliability engineering [13]. The procedures of a single-objective DE (SODE) are presented as follows:

Algorithm 9.4 Differential Evolution

1. Set $\boldsymbol{t} \leftarrow 0$
2. Initialize $\boldsymbol{X_t}$
3. Evaluate $\boldsymbol{X_t}$
4. For $(\boldsymbol{t} \leftarrow \boldsymbol{t}+1, \boldsymbol{t} < \boldsymbol{T_{max}})$
5. Mutate $\boldsymbol{X_t}$ to create $\boldsymbol{V_t}$
6. Crossover $\boldsymbol{V_t}$ to create $\boldsymbol{U_t}$
7. Select $\boldsymbol{X_{t+1}}$ from $\boldsymbol{U_t}$
8. End-for

$\boldsymbol{X}_t = \{\boldsymbol{x}_{1,t}, \ldots, \boldsymbol{x}_{N,t}\}$. Each element $x_{ij,t}$ of the individual $x_{i,t}$ takes a continuous value within its limits: $x_{j_{min}}$ and $x_{j_{max}}$. In Step 5, the mutation procedure is performed as follows. Create N donor vectors each defined as

$$\boldsymbol{v}_{i,t} = \boldsymbol{x}_{r_1,t} + F \cdot \left(\boldsymbol{x}_{r_2,t} - \boldsymbol{x}_{r_3,t}\right), i = 1, \ldots, N \tag{9.3}$$

where F is the scaling factor and $\boldsymbol{x}_{r_1,t}$, $\boldsymbol{x}_{r_2,t}$, and $\boldsymbol{x}_{r_3,t}$ are three randomly chosen individuals in $\boldsymbol{X}_t$,with indexes r_1, r_2, $r_3 \in \{1,2,\ldots,N\}$ and satisfying $r_1 \neq r_2 \neq r_3 \neq i$.

In Step 6, the crossover procedures are presented as follows: Apply the operator in Equation (9.4) to mix each donor vector $\boldsymbol{v}_{i,t}$ with its target vector $\boldsymbol{x}_{i,t}$ to create the corresponding trial vector $\boldsymbol{u}_{i,t}$:

$$u_{ij,t} = \begin{cases} v_{ij,t} & rand_j \leq P_c \text{ or } j = j_{rand} \\ x_{ij,t} & otherwise \end{cases}, \ j = 1, \ldots, K \tag{9.4}$$

where $rand_j$ is a uniform random value lying in the range $[0,1]$, P_c is the crossover probability, K is the vector dimension, and $j_{rand} \in \{1,2,\ldots,K\}$ is a randomly chosen parameter index ensuring $\boldsymbol{u}_{i,t}$ differs from $\boldsymbol{x}_{i,t}$. Evaluate all the trial vectors.

In Step 7, the new population $\boldsymbol{X}_{t+1}$ is generated by selecting the survivors between the target vectors and the corresponding trial vectors, with respect to their fitness values f; in the case of minimization, it proceeds as follows:

$$\boldsymbol{x}_{i,t+1} = \begin{cases} \boldsymbol{u}_{i,t} & f(\boldsymbol{u}_{i,t}) \le f(\boldsymbol{x}_{i,t}) \\ \boldsymbol{x}_{i,t} & f(\boldsymbol{u}_{i,t}) > f(\boldsymbol{x}_{i,t}) \end{cases}, \quad j = 1,\ldots,K \tag{9.5}$$

PSO is based on the social behavior of biological organisms that move in groups, such as birds and fish [14]. Its basic element is a particle representing one candidate solution in the search space. PSO has certain similarities to other EAs, e.g., GAs whereas it is uniquely characterized by the cooperative mechanism. More specifically, all particles change position over time with their own information and that provided by their neighborhoods. One particle's successful adaptation is shared and reflected in the performance of its neighbors [15]. Depending on how the neighborhood is determined, the PSO algorithm may embody the *gbest* and *lbest* models [16]. In the former, each particle interconnects to all others in the swarm, thus sharing information within the whole group. In the latter, a particle does not communicate with the entire swarm of particles but only with some selected ones.

Mathematically, in PSO, a particle i is characterized by three vectors: $\boldsymbol{x}_i^t = (x_{i1}^t,\ldots,x_{in}^t)$, its position in the n-dimensional search space at time t, $\boldsymbol{p}_i = (p_{i1},\ldots,p_{in})$, the best individual position it has thus far visited, and $\boldsymbol{v}_i^t = (v_{i1}^t,\ldots,v_{in}^t)$,its velocity of motion. The procedures of single-objective PSO (SOPSO) with *gbest* model are presented as follows:

Algorithm 9.5 Particle Swarm Optimization

1. Set $\boldsymbol{t} \leftarrow 0$
2. Initialize $\boldsymbol{X_t}$, $\boldsymbol{V_t}$,$\boldsymbol{P}$
3. Evaluate $\boldsymbol{X_t}$
4. Initialize $\mathbf{g} = \mathbf{argmin}_{\boldsymbol{x_i^t} \in \boldsymbol{X_t}} f(\boldsymbol{x_i^t})$
5. For $(\boldsymbol{t} \leftarrow \boldsymbol{t}+1,\ \boldsymbol{t} < \boldsymbol{T_{max}})$
6. Compute $\boldsymbol{X}_{t+1}$ and $\boldsymbol{V}_{t+1}$
7. Evaluate $\boldsymbol{X}_{t+1}$
8. Update $\boldsymbol{P}$ and $\mathbf{g}$
9. End-for

In Step 2: $\boldsymbol{X}_t = \{\boldsymbol{x}_1^t,\ldots,\boldsymbol{x}_N^t\}$ where each particle i is within the range $[\boldsymbol{l}_{xi}, \boldsymbol{u}_{xi}]$; $\boldsymbol{V}_t = \{\boldsymbol{v}_1^t,\ldots,\boldsymbol{v}_N^t\}$, and for each particle i, its velocity vector $\boldsymbol{v}_i^t$ is randomly generated within the range $[\boldsymbol{l}_{vi}, \boldsymbol{u}_{vi}]$; $\boldsymbol{P} = \{\boldsymbol{p}_1,\ldots,\boldsymbol{p}_N\}$, and for each particle i, its best known position is $\boldsymbol{p}_i = \boldsymbol{x}_i^t$. In Step 3, $\mathbf{g}$ is the swarm's best-known position.

In Step 6, for each particle i, its position in the next iteration is computed using the following formulas:

$$\mathbf{x}_i^{t+1} = \mathbf{x}_i^t + \mathbf{v}_i^{t+1} \tag{9.6}$$

$$\mathbf{v}_i^{t+1} = w\mathbf{v}_i^t + c_1 r_1 \left(\mathbf{p}_i - \mathbf{x}_i^t\right) + c_2 r_2 \left(\mathbf{g} - \mathbf{x}_i^t\right) \tag{9.7}$$

where w is the inertia weight determining the exploration scope of the search space; c_1 and c_2 are the acceleration constants for $\mathbf{p}_i$ and $\mathbf{g}$ respectively; r_1 and r_2 are the independent uniform random numbers between 0 and 1.

In Step 8, for each particle i, if $f\left(\mathbf{x}_i^{t+1}\right) < f\left(\mathbf{p}_i\right)$, then set $\mathbf{p}_i = \mathbf{x}_i^{t+1}$; if $f\left(\mathbf{x}_i^{t+1}\right) < f\left(\mathbf{g}\right)$, then set $\mathbf{g} = \mathbf{x}_i^{t+1}$.

9.4 Exercises

1) Write the lines of MATLAB/PYTHON code to simulate the sum distribution of two random variables v and u. If v and u are uniformly distributed in [0, 1], what are the distributions of u+v, u-v, and uv?

2) Write one EA to solve the following redundancy allocation problem, given the parameters in the table below.

$$\max R_s(\mathbf{x}) = \prod_{j=1}^{4}\left[1-\left(1-r_j\right)^{x_j}\right]$$

$$\text{s.t.} \sum_{j=1}^{4} c_{ij} x_j \le b_i, i = 1,2$$

$$1 \le x_j,\ j = 1,2,3,4$$

Stage, j	1	2	3	4
r_j	0.8	0.7	0.75	0.85
c_{1j}	1.2	2.3	3.4	4.5
c_{2j}	5	4	8	7
$b_1 = 56$				
$b_2 = 120$				

3) Assume we have the following function

$$f(x) = x^3 - 60x^2 + 900x + 100$$

where X is constrained to the range of integers [0,..., 31]. We wish to maximize f(X) (the optimal is for X = 10) using a GA. Use a binary representation to represent x by five binary digits.

a) Given the following, four chromosomes give the values for X and f(X).

Chromosome	Binary String
P_1	11100
P_2	01111
P_3	10111
P_4	00100

b) If P_2 and P_3 are chosen as parents and we apply one-point crossover, show the resulting children, C_1 and C_2. Use a crossover point of 1 (where 0 is to the very left of the chromosome).
Do the same using P_2 and P_4 with a crossover point of 2, and create C_3 and C_4.

c) Calculate the value of X and f(X) for C_1 to C_4.

d) Assume that the initial population is $X = \{17, 21, 4, 28\}$. Using one-point crossover, what is the probability of finding the optimal solution? Explain your reasons.

References

1 De Jong, K.A. (2006). *Evolutionary Computation: A Unified Approach*. Cambridge, MA: MIT Press.

2 Bäck, T., Fogel, D.B., and Michalewicz, Z. (1997). *Handbook of Evolutionary Computation*. Boca Raton, FL: CRC Press.

3 Droste, S., Jansen, T., and Wegener, I. (2002). On the analysis of the (1+1) evolutionary algorithm. *Theoretical Computer Science* 276 (1-2): 51–81.

4 Holand, J.H. (1975). *Adaptation in Natural and Artificial Systems*. Ann Arbor: The University of Michigan Press.

5 Miller, B.L. and Goldberg, D.E. (1995). Genetic algorithms, tournament selection, and the effects of noise. *Complex Systems* 9 (3): 193–212.

6 Goldberg, D.E. and Deb, K. (1991). A comparative analysis of selection schemes used in genetic algorithms. *Foundations of Genetic Algorithms*, 1, 69–93.

7 Eiben, A.E. and Smith, J.E. (2003). *Introduction to Evolutionary Computing*. New York City: Springer.

8 Koza, J.R. (1992). *Genetic Programming: On the Programming of Computers by Means of Natural Selection*. Cambridge, MA: MIT Press.

9 De Jong, K.A. and Spears, W.M. (1992). A formal analysis of the role of multi-point crossover in genetic algorithms. *Annals of Mathematics and Artificial Intelligence* 5 (1): 1–26.

10 Herrera, F., Lozano, M., and Verdegay, J.L. (1998). Tackling real-coded genetic algorithms: Operators and tools for behavioural analysis. *Artificial Intelligence Review* 12 (4): 265–319.

11 Rudolph, G. (1994). Convergence analysis of canonical genetic algorithms. *IEEE Transactions on Neural Networks* 5 (1): 96–101.

12 Storn, R. and Price, K. (1995). *Differential Evolution-a Simple and Efficient Adaptive Scheme for Global Optimization over Continuous Spaces*, Vols. TR-95-012. Berkeley, CA: International Computer Science Institute.

13 Arya, L.D.A.L.D., Choube, S.C., and Arya, R. (Feb 2011). Differential evolution applied for reliability optimization of radial distribution systems. *International Journal of Electrical Power & Energy Systems* 33 (2): 271–277.

14 Kennedy, J. and Eberhart, R. (1995). Particle swarm optimization. in *Proceedings of 1995 IEEE International Conference on Neural Networks*, pp. 1942–1948.

15 Kennedy, J., Kennedy, J.F., and Eberhart, R.C. (2001). *Swarm Intelligence*. San Francisco: USA Morgan Kaufmann.

16 Bratton, D. and Kennedy, J., "Defining a standard for particle swarm optimization," in *Swarm Intelligence Symposium, 2007. SIS 2007. IEEE*, Honolulu, Hawaii, 2007: IEEE, pp. 120–127.

10

Multi-Objective Optimization (MOO)

System reliability optimization typically considers multiple reliability-related objectives: reliability, availability, and maintainability (RAM). For hazardous systems, risk attributes must also be considered, i.e. consideration of RAM and Safety criteria (RAMS) [1]. Moreover, any design, inspection, and maintenance activity is associated with a cost. In conclusion, system reliability optimization has essentially a multi-objective formulation, which aims at finding the appropriate choices of reliability design, inspection and maintenance procedures that optimally balance the conflicting RAMS and cost attributes (RAMS+C) [2].

Then the decision variable vector $\boldsymbol{x}$ is evaluated with respect to multiple numerical objectives related to the RAMS+C attributes: $R(\boldsymbol{x})$ = system reliability; $A(\boldsymbol{x})$ = system availability; $M(\boldsymbol{x})$ = system maintainability, e.g. the unavailability contribution due to failures but also test and maintenance; $S(\boldsymbol{x})$ = system safety, normally quantified in terms of the system risk measure $Risk(\boldsymbol{x})$ (e.g. as assessed from a probabilistic risk analysis); and $C(\boldsymbol{x})$ = cost required to implement the vector choice $\boldsymbol{x}$. Many works convert the multi-objective optimization (MOO) problem into a single-objective one by, e.g. regarding one of the RAMS attributes or cost as the single objective and the other attributes as constraints or by aggregating all attributes into a single objective. Then the techniques of the solution to single-objective optimization (SOO) problems that have been documented in Chapters 8 and 9 can be used.

This chapter mainly focuses on MOO problems. It covers various topics related to MOO, including MOO problem formulation, method of conversion from MOO problem to SOO problem, MOO evolutionary algorithms, performance measures and method of selection of the preferred solutions. Finally, the guidelines of implementing and developing MOO methods for solving RAMS+C problems are presented.

10.1 Multi-objective Problem Formulation

In general, a MOO (minimization) problem can be formulated as follows:

$$\min f_i(x), i = 1, \ldots, M \tag{10.1a}$$

Reliability Analysis, Safety Assessment and Optimization: Methods and Applications in Energy Systems and Other Applications, First Edition. Enrico Zio and Yan-Fu Li.

$$\text{s.t.}\begin{cases} g_j(x)=0, & j=1,\ldots,J \\ h_k(x)\leq 0, & k=1,\ldots,K \end{cases} \tag{10.1b}$$

where f_i is the i-th of the M objective functions, $\boldsymbol{x}=(x_1,x_2,\ldots,x_N)$ is the decision variable vector that represents a solution in the solution space $\mathbf{R}^N$, g_j is the j-th of the J equality constraints, and h_k is the k-th of the K inequality constraints. Let $z_i=f_i(x),\forall i$; then $z=(z_1,z_2,\ldots,z_M)$ is the objective vector and z is inside $\mathbf{R}^M$, the objective space.

For ease of notation, we assume all objective functions are to be minimized: If any $f_i(x)$ were to be maximized, they can be converted into $1-f_i(x)$ for minimization. Adopting the general definition for RAMS+C optimization, the MOO problem has the following formulation:

$$\min\left(1-R(\boldsymbol{x}),1-A(\boldsymbol{x}),M(\boldsymbol{x}),Risk(\boldsymbol{x}),C(\boldsymbol{x})\right) \tag{10.2a}$$

$$\text{s.t.}R(\boldsymbol{x})\geq R_L \tag{10.2b}$$

$$A(\boldsymbol{x})\geq A_L \tag{10.2d}$$

$$M(\boldsymbol{x})\leq M_U \tag{10.2e}$$

$$Risk(\boldsymbol{x})\leq R_U \tag{10.2f}$$

$$C(\boldsymbol{x})\leq C_U \tag{10.2g}$$

$$\boldsymbol{x}=\left(\boldsymbol{x}_1,\ldots,\boldsymbol{x}_{N_d}\right)\in\mathbf{R}^{N_d} \tag{10.2h}$$

The quantities R_L,A_L,M_U,R_U,C_U represent the constraining threshold values for the reliability, unavailability, maintainability, risk, and cost objectives, respectively. As mentioned in Chapter 1, *Reliability* of a certain component or system measures its capability to sustain operation without failure under specified conditions during a given period of time. It is an intrinsic property that directly depends on the component's or system's physical characteristics and its design, rather than on its maintenance. Maintenance, on the other hand, relates to all activities performed on the component or system during the operational lifetime to sustain or restore its functional capabilities. In spite of its positive effects on component or system functionality, maintenance activities could result to the downtime of the component or system during which the system might not perform its designated functions. *Availability* measures the probability that the component or system performs its designated functions at any time point considering unplanned failure interruptions and planned maintenance activities. *Maintainability* measures the capability of the system to be maintained under specified conditions during a given period of time. For quantitative analysis, the above mentioned metrics are typically all defined in probabilistic terms. *Safety* is defined as the capability to prevent or mitigate the consequences of postulated accidents on specified targets (e.g. workers, public, and environment); risk is often adopted as the quantitative metric of interest, in relation to scenarios, probabilities of occurrence, and consequences.

MOO requires minimizing all objectives simultaneously. If no conflict exists between any pair of the objectives, one would find a single solution that minimizes all objectives at the same time. In this case, solving the MOO problem is equivalent to minimizing one of the objectives. The MOO methods need to be applied only when conflicts exist among the objectives. In this case, due to the contradiction and possible incommensurability of the objective functions, MOO methods identify a set of optimal solutions $\boldsymbol{x}_l^*, l = 1,2,\ldots,L$ instead of a single optimal solution.

In the set of optimal solutions of a MOO problem, no one can be regarded as better than any other with respect to all the objective functions. The identification of this set of solutions can be achieved in terms of the concepts of Pareto optimality and dominance [3]. In case of a minimization problem, the solution $\boldsymbol{x}_a$ is regarded to dominate solution $\boldsymbol{x}_b$ ($\boldsymbol{x}_a \succ \boldsymbol{x}_b$) if both following conditions are satisfied:

$$\forall i \in \{1,2,\ldots,M\}, f_i(\boldsymbol{x}_a) \leq f_i(\boldsymbol{x}_b) \tag{10.3a}$$

$$\exists j \in \{1,2,\ldots,M\}, f_j(\boldsymbol{x}_a) < f_j(\boldsymbol{x}_b) \tag{10.3b}$$

If one or both of the above conditions are violated, $\boldsymbol{x}_b$ is said to be non-dominated by $\boldsymbol{x}_a$. Within the entire search space, the solutions non-dominated by any others are *Pareto-optimal* and constitute the *Pareto-optimal set*; the corresponding $\boldsymbol{z}$ objective functions values form the *Pareto-optimal front* in $\mathbf{R}^M$. The goal of a MOO method is to search for solutions in the Pareto-optimal set while maintaining diversity so as to cover the Pareto-optimal front. Therefore, flexibility is allowed in the final decisions on the solutions to be implemented (Figure 10.1).

10.2 MOO-to-SOO Problem Conversion Methods

The goal of the methods for MOO problem solution is to obtain the Pareto-optimal front, or Pareto front in short. There are typically two ways of achieving this goal. The first one is to convert the MOO problem into multiple SOO problems, such that the solution to each SOO problem produces one member of the Pareto-optimal front. The second one is to simultaneously optimize the multiple objectives. In this section, the MOO-SOO conversion methods are introduced: once the SOO problems are obtained, then the methods introduced in Chapters 8 and 9 can be applied.

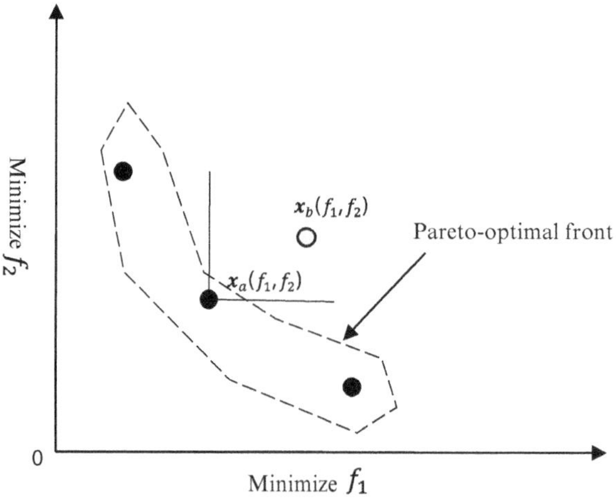

Figure 10.1 Pareto dominance and Pareto optimality.

10.2.1 Weighted-sum Approach

The weighted-sum approach aggregates multiple objectives into one single objective using a weighting vector $\boldsymbol{w} = (w_1, w_2, \ldots, w_M)$ as in the following:

$$\min \sum_{i=1}^{M} w_i f_i(\boldsymbol{x}) \tag{10.4a}$$

$$\text{s.t.} \begin{cases} g_j(\boldsymbol{x}) = 0, & j = 1, \ldots, J \\ h_k(\boldsymbol{x}) \leq 0, & k = 1, \ldots, K \end{cases} \tag{10.4b}$$

where each $w_i \geq 0$ and $\sum_{i=1}^{M} w_i = 1$. Consequently, the solution to Equation (10.4) is also a solution to Equation (10.1). The coefficients are selected depending on the decision maker (DM) preferences. In multiple criteria decision analysis theory, there are a number of methods, e.g. AHP and TOPSIS, that have been developed for the quantification of the DMs preferences. Interested readers could refer to [4] for detailed information.

This approach is most straightforward to convert a MOO problem. For problems that have a convex Pareto-optimal front, it guarantees finding the solutions on the entire Pareto-optimal set by varying the values of the weight vector. However, this method has four disadvantages: Different weight vectors do not necessarily result into different Pareto-optimal solutions; uniformly distributed sets of weight vectors do not necessarily result into uniformly distributed Pareto-optimal solutions; there are difficulties to handle objectives of different numerical scales; and there are difficulties to find certain Pareto-optimal solutions in a non-convex objective space.

10.2.2 ε-constraint Approach

To alleviate the difficulties faced by the weighted-sum approach, the ε-constraint approach was proposed in 1971 by Haimes, et al. [5]. It reformulates the MOO problem by keeping one objective and transforming the others into constraints bounded by user-specific values. The transformed problem has the following expression:

$$\text{Min } f_\theta(\boldsymbol{x}), \tag{10.5a}$$

$$\text{s.t.}\begin{cases} f_i(\boldsymbol{x}) \le \varepsilon_i, & i = 1,\ldots,M \ and\ i \ne \theta \\ g_j(\boldsymbol{x}) = 0, & j = 1,\ldots,J \\ h_k(\boldsymbol{x}) \le 0, & k = 1,\ldots,K \end{cases} \tag{10.5b}$$

where ε_i is the upper bound of $f_i(\boldsymbol{x})$ and satisfies that $L_i \le \varepsilon_i \le U_i$. By iteratively increasing or decreasing the value of ε_i, in principle, we can obtain all Pareto-optimal solutions on the entire Pareto-optimal set regardless the convexity of the Pareto front. The disadvantages of this method lie in the difficulties to determine the ranges of the objectives being constrained and the values of ε_i, especially when many objectives are involved. The lower limit of each converted objective L_i can be obtained by solving the individual SOO using the same objective. However, computing the upper limit U_i is not straightforward. A Payoff Table (as shown in Table 10.1) is usually implemented [6]. In this table, the leftmost cell of each row contains the optimal solution to an individual objective, e.g. $\boldsymbol{x}_i^{min}$ represents the optimal solution to the ith objective. The cell of row i under column j includes the jth objective function value, given $\boldsymbol{x}_i^{min}$. Through this table, the upper limit U_i can be estimated by $\max_j\left\{f_i\left(\boldsymbol{x}_j^{min}\right)\right\}$. However, there can be large discrepancy between this estimation and the real upper limit. The readers can refer to [6] for details about finding the real upper limit.

10.2.3 Goal Programming

Goal programming was first introduced in 1955 by Charnes, et al. [7] to solve a single-objective linear programming (LP) problem and is widely used for solving MOO problems. The main idea of goal programming is to find the solutions that achieve the

Table 10.1 Payoff Table of the MOO problem formulation.

	z_1	...	z_M
$\boldsymbol{x}_1^{min}$	$f_1\left(\boldsymbol{x}_1^{min}\right)$	...	$f_M\left(\boldsymbol{x}_1^{min}\right)$
...	...		...
$\boldsymbol{x}_M^{min}$	$f_1\left(\boldsymbol{x}_M^{min}\right)$	...	$f_M\left(\boldsymbol{x}_M^{min}\right)$

predefined targets at one or more objectives. If there is no solution to achieving the predefined targets in all objectives, the task will be to find the solutions that minimize the deviations from the objectives. On the other hand, if solutions exist within the desired targets, the task will be to identify those solutions.

For a MOO problem, the simplest version of goal programming requires the DM to set the target and relative weight for each objective function. An optimal solution $\boldsymbol{x}^*$ is defined as the one that minimizes the deviation from the set targets. Goal programming generally takes the following form:

$$\text{Min} \sum_{i=1}^{M} c_i \left(d_i^+ + d_i^- \right) \tag{10.6a}$$

$$\text{s.t.} \begin{cases} f_i(\boldsymbol{x}) + d_i^+ - d_i^- = f_i^0 \\ d_i^+ d_i^- = 0 \\ d_i^+, d_i^- \geq 0 \qquad i = 1, \ldots, M \end{cases} \tag{10.6b}$$

where c_i is the weight of the deviation of each objective, d_i^+ and d_i^- are respectively the positive and negative deviations, and f_i^0 is the predefined target for the i-th objective. The disadvantages of this method are similar to the weighted sum approach, as the DM has to provide targets and weights for each of the objective functions.

10.3 Multi-objective Evolutionary Algorithms

The above mentioned approaches for solving MOO problems are often referred to as classical. They all suggest certain ways of converting a MOO problem into a SOO problem. They have some common difficulties, such as that only one Pareto-optimal solution can be found in one simulation run and that certain problem knowledge, such as w_i, ε_i, c_i and f_i^0 is required from the DM.

The EAs, e.g. genetic algorithms (GAs) [8], introduced in Chapter 9, are stochastic optimization methods mimicking biological evolution on a group of individuals (solutions). The parallelization and evolution operations of EAs are well-suited to the characteristics of MOO problems. Parallelization helps to identify multiple solutions on the Pareto front in one run without soliciting knowledge from the DM; the evolution operators have the capability to avoid trapping into the local minima (which is common in non-convex objective spaces). These properties render the EAs by far the most popular methods implemented for RAMS+C MOO. There are several EAs specifically developed for solving MOO problems. In the following, we will introduce two representative ones.

10.3.1 Fast Non-dominated Sorting Genetic Algorithm (NSGA-II)

Fast non-dominated sorting genetic algorithm (NSGA-II) [9] has become one of the standard approaches of multi-objective EAs (MOEAs). The input parameters are N

population size, P_c crossover probability, P_m mutation probability and T maximum number of generations. The output is $\boldsymbol{P}_T$ final population. The procedure of NSGA-II is presented as the following:

Procedure 10.1 Nsga-II

1. Initialization: Set the generation counter $t = 0$; randomly generate an initial population $\boldsymbol{P}_t$ of size N.
2. Mating selection: Perform the binary tournament selection with replacement on $\boldsymbol{P}_t$ to select parents to be processed by genetic operators.
3. Variation: Apply crossover and mutation operators to the paired parents with probability $\boldsymbol{P}_c$ and $\boldsymbol{P}_m$, respectively, to create offspring population Q_t of size N.
4. Dominance ranking: $\boldsymbol{F} = \boldsymbol{P}_t \cup \boldsymbol{Q}_t$, then use the fast non-dominated sorting algorithm to identify the non-dominated fronts $\boldsymbol{F}_1, \boldsymbol{F}_2, \ldots, \boldsymbol{F}_k$ in the union $\boldsymbol{F}$.
5. Environmental selection:
   - 5.1 Set $\boldsymbol{P}_{t+1} = \varnothing$, then perform what follows;
   - 5.2 For $i = 1, \ldots, k$ do the following steps;
   - 5.3 If $|\boldsymbol{P}_{t+1}| + |\boldsymbol{F}_i| \leq N$, then set $\boldsymbol{P}_{t+1} = \boldsymbol{P}_{t+1} \cup \boldsymbol{F}_i$; and
   - 5.4 Else, calculate crowding distance of the solutions in $\boldsymbol{F}_i$; add the least crowded $N - |P_{t+1}|$ solutions of $\boldsymbol{F}_i$ to $\boldsymbol{P}_{t+1}$.
6. Set $t = t + 1$.
7. Termination: $t > T$, then stop and return $\boldsymbol{P}_t$; otherwise, go to Step 2.

The procedures above show three key concepts: dominance ranking, fast non-dominated sorting algorithm, and crowding distance. In the fast non-dominated sorting algorithm, for each solution, there are two entities: domination count n_x, i.e. the number of solutions which dominate the present solution $\boldsymbol{x}$ and $\boldsymbol{S}_x$, a set of solution that the solution $\boldsymbol{x}$ dominates. The algorithm is presented as follows:

Algorithm 10.1 Fast non-dominated sorting.

1. For each $\boldsymbol{x} \in \boldsymbol{P}$
2.   Set $\boldsymbol{n}_x \leftarrow 0$, $\boldsymbol{S}_x \leftarrow \varnothing$
3.   For each $\boldsymbol{x}' \in \boldsymbol{P}$
4.   If $\boldsymbol{x} \succ \boldsymbol{x}'$ then // if $\boldsymbol{x}$ dominates $\boldsymbol{x}'$
5. $\boldsymbol{S}_x \leftarrow \boldsymbol{S}_x \cup \{\boldsymbol{x}'\}$ // add $\boldsymbol{x}'$ to the set of solutions dominated by $\boldsymbol{x}$
6.   Else if $\boldsymbol{x}' \succ \boldsymbol{x}$ then
7. $n_x \leftarrow n_x + 1$ // increase the domination counter of $\boldsymbol{x}$
8.   End-if
9.   End-for
10.   If $n_x = 0$ then // $\boldsymbol{x}$ belongs to the first front
11. $\boldsymbol{x}_{rank} \leftarrow 1$ // assign front number to $\boldsymbol{x}$

12. $F_1 \leftarrow F_1 \cup \{x\}$
13.   End-if
14. End-for
15. Set $k \leftarrow 1$ // initialize the front counter
16. While $F_k \neq \emptyset$
17. $Q \leftarrow \emptyset$
18.   For each $x \in F_k$
19.   For each $x' \in S_x$
20. $n_{x'} \leftarrow n_{x'} - 1$
21.   If $n_{x'} = 0$ then // x' belongs to the next front
22. $x'_{rank} \leftarrow k + 1$ // assign front number to x'
23. $Q \leftarrow Q \cup \{x'\}$
24.   End-if
25.   End-for
26.   End-for
27. $k \leftarrow k + 1$
28. $F_k \leftarrow Q$
29. End-while

where x is the index of x in the current population. The outputs of this algorithm include the total number of front k and all non-dominated fronts $F_1, F_2, ..., F_k$. Figure 10.2 shows an example of this sorting.

To identify the different Pareto-optimal fronts in Figure 10.2, the step are shown in Table 10.2 according to the fast non-dominated sorting algorithm, i.e. Algorithm 10.1.

Crowding distance is another important concept in NSGA-II. Suppose there are $l = |F|$ solutions on the front. For each objective function k, sort the l solutions in worsening order, and let $x_{[i,k]}$ represent the i-th solution in the sorted list with respect to objective function k. The definition of crowding distance is shown as follows:

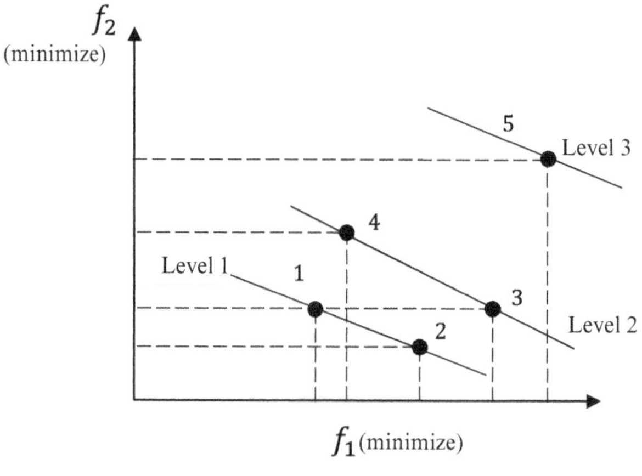

Figure 10.2 Fast non-dominance sorting, an example.

Table 10.2 Steps of fast non-dominance sorting, an example.

1	For each $\boldsymbol{x} \in \mathbf{P}$, set $n_{\boldsymbol{x}} = 0$ and $S_{\boldsymbol{x}} = \phi$.
2	For solution 1, since $1 \succ 3$, $1 \succ 4$ and $1 \succ 5$, update $\boldsymbol{S}_1 = \{3,4,5\}$ and $n_1 = 0$. For solution 2, since $2 \succ 3$, $2 \succ 4$ and $2 \succ 5$, update $\boldsymbol{S}_2 = \{3,4,5\}$ and $n_2 = 0$. For solution 3, since $3 \succ 5$, $2 \succ 3$ and $1 \succ 3$, update $\boldsymbol{S}_3 = \{5\}$ and $n_3 = 2$. For solution 4, since $4 \succ 5$ and $1 \succ 4$, update $\boldsymbol{S}_4 = \{5\}$ and $n_4 = 1$. For solution 5, since $1 \succ 5$, $2 \succ 5$, $3 \succ 5$ and $4 \succ 5$, update $\boldsymbol{S}_5 = \phi$ and $n_5 = 4$.
3	Since $n_1 = n_2 = 0$, set $\boldsymbol{F}_1 = \{1,2\}$. Set $k = 1$.
4	For $x' \in \boldsymbol{S}_1$, update $n_{x'} = n_{x'} - 1$. For $\boldsymbol{x}' \in \boldsymbol{S}_2$, update $n_{x'} = n_{x'} - 1$. Obtain $n_3 = 0$, $n_3 = 0$ and $n_5 = 2$.
5	Set $k = 2$. Since $n_3 = n_4 = 0$, set $F_2 = \{3,4\}$.
6	For $x' \in \boldsymbol{S}_3$, update $n_{x'} = n_{x'} - 1$. For $x' \in \boldsymbol{S}_4$, update $n_{x'} = n_{x'} - 1$. Obtain $n_5 = 0$.
7	Set $k = 3$. Since $n_5 = 0$, set $F_3 = \{5\}$.
8	Stop and declare the total number of fronts $k = 3$ and all non-dominated sets $\boldsymbol{F}_i$, for $i = 1,2,3$.

$$d_{\boldsymbol{x}_{[i,k]}}^k = \frac{f_k\left(\boldsymbol{x}_{[i+1,k]}\right) - f_k\left(\boldsymbol{x}_{[i-1,k]}\right)}{f_k^{\max} - f_k^{\min}} \text{for}\, i = 2,\ldots,l-1 \tag{10.7}$$

We have the two extreme cases: $d_{\boldsymbol{x}_{[1,k]}}^k = \infty$ and $d_{\boldsymbol{x}_{[l,k]}}^k = \infty$. For each solution $\boldsymbol{x}$, the crowding distance is the sum of all its crowding distances, each with respect to one objective, i.e. $d_{\boldsymbol{x}} = \sum_k d_{\boldsymbol{x}}^k$.

Take solution 4 in Figure 10.3 for example. For f_1, sort solutions as $\{1,2,3,4,5,6\}$ and $d_4^1 = \frac{f_1(\boldsymbol{x}_5) - f_1(\boldsymbol{x}_3)}{f_1^{\max} - f_1^{\min}}$. For f_2, sort solutions as $\{6,5,4,3,2,1\}$ and $d_4^2 = \frac{f_2(\boldsymbol{x}_3) - f_2(\boldsymbol{x}_5)}{f_2^{\max} - f_2^{\min}}$. The crowding distance of 4 is $d_4 = d_4^1 + d_4^2$.

NSGA-II has three major advantages: $O\left(MN^2\right)$ computational complexity of sorting (where M is the number of objectives and N is the population size), elitism approach, and self-maintained diversity. In theory, NSGA-II is a GA without elitism strategy because no mechanism preserves the best solutions found in each generation.

10.3.2 Improved Strength Pareto Evolutionary Algorithm (SPEA 2)

The strength Pareto evolutionary algorithm (SPEA) [10] is an elitist EA. The elitism is introduced by explicitly maintaining an external archive $\boldsymbol{E}_t$ of non-dominated solutions

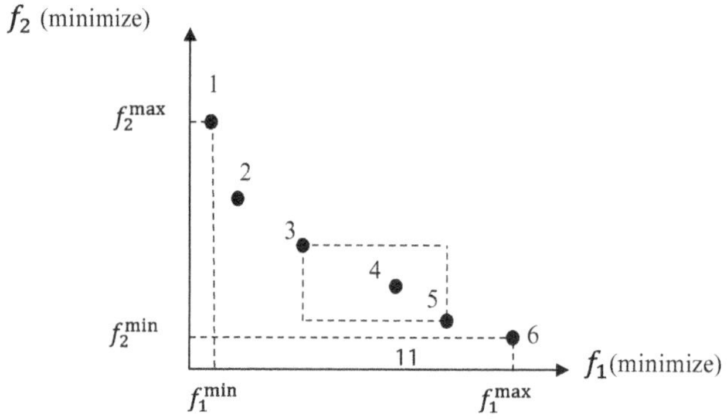

Figure 10.3 An example of crowding distance.

in the course of searching. It is able to retain the elites using the ranking principle in the environmental selection step and is characterized by the clustering mechanism to truncate the external population to increase the diversity of non-dominated solutions. Despite the advantages mentioned above, SPEA is typically time-consuming, mainly due to the complexity of the clustering algorithm. Thus, the improved SPEA (SPEA 2) was proposed for better performance [11]. Different from SPEA, SPEA 2 considers the domination strength of each solution and applies the k-th nearest neighbor-based density to maintain population diversity. In SPEA 2, the fitness assignment incorporating domination strength and density information is the diversity-preserving mechanism.
The procedures of SPEA 2 are presented as follows.

Procedure 10.2 SPEA 2

1. Initialization: Set the generation counter $t = 0$; randomly generate an initial population $\boldsymbol{P}_t$ of size N; create the empty archive (external population) $\boldsymbol{E}_t$ of size N_E.
2. Fitness assignment: Calculate the fitness of each solution $\boldsymbol{x}$ in $\boldsymbol{P}_t \cup \boldsymbol{E}_t$ via the following steps:

   2.1. Compute the raw fitness of solution $\boldsymbol{x} : R(\boldsymbol{x}, t) = \sum_{\boldsymbol{y} \in \boldsymbol{P}_t \cup \boldsymbol{E}_t, \boldsymbol{y} \succ \boldsymbol{x}} S(\boldsymbol{y}, t)$ where $S(\boldsymbol{y}, t)$ is the number of solutions in $\boldsymbol{P}_t \cup \boldsymbol{E}_t$ dominated by the solution $\boldsymbol{y}$, i.e. the strength of $\boldsymbol{y}$. For a non-dominated solution $\boldsymbol{x}$, set $R(\boldsymbol{x}, t) = 0$.

   2.2. Calculate the density estimate of solution $\boldsymbol{x}$: $D(\boldsymbol{x}, t) = \left(\sigma_x^k + 2\right)^{-1}$ where $\boldsymbol{\sigma}_x^k$ is the distance between $\boldsymbol{x}$ and its k-th nearest neighbor. A common setting for k is $\sqrt{N + N_E}$.

   2.3. Assign the fitness value to solution $\boldsymbol{x}$: $F(\boldsymbol{x}, \boldsymbol{t}) = R(\boldsymbol{x}, \boldsymbol{t}) + D(\boldsymbol{x}, \boldsymbol{t})$.

3. Environmental selection: Copy all non-dominated solutions, the fitness values of which are lower than one, from $\boldsymbol{P}_t \cup \boldsymbol{E}_t$ to $\boldsymbol{E}_{t+1}$.

3.1. If $|\boldsymbol{E}_{t+1}| \le N_E$, then add the best $N_E - |\boldsymbol{E}_{t+1}|$ dominated solutions of $\boldsymbol{P}_t \cup \boldsymbol{E}_t$ into $\boldsymbol{E}_{t+1}$ according to the fitness values;
3.2. Else, then iteratively remove $|\boldsymbol{E}_{t+1}| - N_E$ solutions with respect to density. Break any tie by examining the maximum σ^l for $l = k-1,\ldots,1$, sequentially.

4. Termination: If the stopping criterion is satisfied, then stop and return the set of non-dominated solutions in $\boldsymbol{E}_{t+1}$.
5. Mating selection: Perform the binary tournament selection with replacement on $\boldsymbol{E}_{t+1}$ to select parents for genetic operators.
6. Variation: Apply crossover and mutation operators to the parents to create offspring solutions which constitute the next generation $\boldsymbol{P}_{t+1}$. Set $t = t+1$ and go to Step 2.

In SPEA 2, there are two key mechanisms: fitness assignment and environmental selection. For the former, we will explain the procedure to compute the final fitness value of a solution $\boldsymbol{x}$. The domination strength $S(\boldsymbol{y})$ is the number of solutions it dominates in archive $\boldsymbol{E}_t$ and population $\boldsymbol{P}_t$. It is mathematically defined as follows:

$$S(\boldsymbol{y}) = \left|\{\boldsymbol{y}' \mid \boldsymbol{y}' \in \boldsymbol{P}_t \cup \boldsymbol{E}_t, \boldsymbol{y} \succ \boldsymbol{y}'\}\right| \tag{10.8}$$

The raw fitness solution $\boldsymbol{x}$ is determined by the strengths of its dominators in archive $\boldsymbol{E}_t$ and population $\boldsymbol{P}_t$. Its definition is shown as follows:

$$R(\boldsymbol{x}) = \sum_{\boldsymbol{y} \in \boldsymbol{P}_t \cup \boldsymbol{E}_t, \boldsymbol{y} \succ \boldsymbol{x}} S(\boldsymbol{y}) \tag{10.9}$$

A high $R(\boldsymbol{x})$ value means that $\boldsymbol{x}$ is dominated by many solutions. For a non-dominated solution $\boldsymbol{x}$, we have $R(\boldsymbol{x}) = 0$. Although the raw fitness assignment provides a type of niching mechanism based on the Pareto dominance concept, it may fail when most individuals do not dominate each other. Thus, additional density information needs to be incorporated to the fitness.

The density estimate $D(\boldsymbol{x})$ for solution $\boldsymbol{x}$ is defined as the inverse of the distance $\sigma_{\boldsymbol{x}}^k$ to its k-th nearest neighbor in archive $\boldsymbol{E}_t$ and population $\boldsymbol{P}_t$:

$$D(\boldsymbol{x}) = \left(\sigma_{\boldsymbol{x}}^k + 2\right)^{-1} \tag{10.10}$$

In the denominator, 2 is added to ensure that $D(\boldsymbol{x}) < 1$. The final fitness value is defined as the sum of the raw fitness and the density estimate:

$$F(\boldsymbol{x}) = R(\boldsymbol{x}) + D(\boldsymbol{x}) \tag{10.11}$$

For a non-dominated solution $\boldsymbol{x}$, its fitness value is $F(\boldsymbol{x}) < 1$; for a dominated solution $\boldsymbol{x}$, its fitness value is $F(\boldsymbol{x}) > 1$.

For the environmental selection, all non-dominated solutions in $P_t \cup E_t$, whose fitness values are lower than one, are first copied into E_{t+1}; then the archive truncation is performed considering the following two cases.

Case 1: If $|E_{t+1}| \le N_E$, add the best $N_E - |E_{t+1}|$ dominated solutions of $P_t \cup E_t$ into E_{t+1} according to the fitness values.

Case 2: If $|E_{t+1}| > N_E$, iteratively remove $|E_{t+1}| - N_E$ solutions with respect to density. At each removal iteration, solution $\boldsymbol{x}$ to be removed should satisfy $\boldsymbol{x} \le_d \boldsymbol{y}$ for all $\boldsymbol{y} \in E_{t+1}$. The relation $\boldsymbol{x} \le_d \boldsymbol{y}$ is defined as

$$\forall 0 < k < |E_{t+1}| : \sigma_{\boldsymbol{x}}^k = \sigma_{\boldsymbol{y}}^k \vee$$

$$\exists 0 < k < |E_{t+1}| : [(\forall 0 < l < k : \sigma_{\boldsymbol{x}}^l = \sigma_{\boldsymbol{y}}^l) \wedge \sigma_{\boldsymbol{x}}^k < \sigma_{\boldsymbol{y}}^k] \tag{10.12}$$

In the following Table 10.3, we show one example of the removal iterations. Given the solution and its objective values in the first three columns, the distance of each solution to all other solutions are calculated and sorted in increasing order in the fourth column. The distances to the first and second nearest neighbors are shown in the fifth and sixth columns. We can see that in iteration #1, solution 4 is removed. Then all the distances are recalculated after the removal. In iteration #2, solution 1 is removed.

To summarize, the SPEA 2 has the following advantages: It provides a better distribution of Pareto-optimal solutions than NSGA-II, especially when the number of objectives increases, and the archive truncation guarantees the preservation of boundary solutions. Its disadvantages mainly lie in the computational complexities, i.e. calculation of density estimator and the calculation of fitness are time-consuming.

Table 10.3 Example of removal iterations.

Iteration #1					
Solution $\boldsymbol{x}$	f_1	f_2	Distance to all solutions in increasing order	σ_x^1	σ_x^2
1	0.31	6.10	[0.83, 0.99, 1.60, 2.47]	0.83	–
2	0.22	7.09	[0.17, 0.99, 2.60, 3.47]	0.17	0.99
3	0.66	3.65	[0.87, 2.47, 3.30, 3.47]	0.87	–
4	0.27	6.93	[0.17, 0.83, 2.43, 3.30]	0.17	0.83
5	0.58	4.52	[0.87, 1.60, 2.43, 2.60]	0.87	–
Iteration #2					
Solution $\boldsymbol{x}$	f_1	f_2	Distance to all solutions in increasing order	σ_x^1	σ_x^2
1	0.31	6.10	[0.70, 1.60, 2.47]	0.70	1.60
2	0.22	7.09	[0.70, 2.28, 3.15]	0.70	2.28
3	0.66	3.65	[0.87, 2.47, 3.15]	0.87	–
5	0.58	4.52	[0.87, 1.60, 2.28]	0.87	–

Besides NSGA-II and SPEA 2, other well-known MOGAs include vector-evaluated GA (VEGA) [12] and niched Pareto GA (NPGA) [13]. The details about MOEAs for RAMS+C optimization have been well-documented in the tutorials by Marseguerra and Zio [2] and, Konak et al. [14]. Despite the popularity of the MOEAs, in general, they have the following disadvantages: There is no guarantee to find a true Pareto-optimal solution; there is no guarantee to identify all Pareto-optimal solutions; they are computationally expensive for large population sizes.

10.4 Performance Measures

The performance of the MOO methods needs to be evaluated quantitatively to guide the creation and implementation of efficient MOO methods for the different problems. Performance measures are defined for this purpose. There are two goals for MOO: discover solutions as close to the Pareto-optimal front as possible (i.e., search for the Pareto-optimal front) and maintain a diverse set of Pareto-optimal solutions (i.e., search along the Pareto-optimal front). A MOO method is considered of good performance if both of the above goals are sufficiently satisfied. Correspondingly, the performance measures are of three categories: measure to evaluate closeness to the true Pareto-optimal front; measure to evaluate diversity among non-dominated solutions; and measure to evaluate closeness and diversity. For more comprehensive information about the performance measures, please refer to [15,16].

In this section, we introduce three representative measures, each belonging to one category. The first measure is named as generational distance (GD) [17]. It explicitly computes the closeness of a non-dominated solution set $\boldsymbol{Q}$ to a known Pareto-optimal set P^*. It is mathematically defined as

$$GD = \frac{\left(\sum_{i=1}^{|Q|} d_i^p\right)^{1/p}}{|Q|} \tag{10.13}$$

For $p = 2$, it defines d_i as the Euclidean distance (in the objective space) between solution $i \in Q$ and its nearest solution in P^*, i.e. $d_i = \min_{k \in P^*} \sqrt{\sum_{m=1}^{M} \left(f_m^{(i)} - f_m^{*\,(k)}\right)^2}$ where $f_m^{*\,(k)}$ denotes the m-th objective function value of the k-th solutions in P^*. Figure 10.4 shows one example of computing GD.

An algorithm having a smaller value of GD is regarded superior to another one having a larger value of GD. The disadvantages of GD as a measure of performance is that the set P^* should be known; otherwise, it is necessary to find an appropriate set, which can be considered as P^* before computing GD.

The second measure is named spacing, which quantifies the diversity of the non-dominated front. It is calculated with a relative distance measure between consecutive solutions in the non-dominated set Q, as in equation (10.14):

Set P^*

Solution	f_1	f_2
1	1.0	7.5
2	1.1	5.5
3	2.0	5.0
4	3.0	4.0
5	4.0	2.8
6	5.5	2.5
7	6.8	2.0
8	8.4	1.2

Set Q

Solution	f_1	f_2
A	1.2	7.8
B	2.8	5.1
C	4.0	2.8
D	7.0	2.2
E	8.4	1.2

Solution	Euclidean distance d_i
A	$\sqrt{(1.2-1.0)^2+(7.8-7.5)^2}=0.36$
B	$\sqrt{(2.8-2.0)^2+(5.1-5.0)^2}=0.81$
C	$\sqrt{(4.0-4.0)^2+(2.8-2.8)^2}=0.00$
D	$\sqrt{(7.0-6.8)^2+(2.2-2.0)^2}=0.28$
E	$\sqrt{(8.4-8.4)^2+(1.2-1.2)^2}=0.00$

$$\text{GD}=\frac{(\sum_{i=1}^{5} d_i^2)^{1/2}}{5}=\frac{(0.36^2+0.81^2+0^2+0.28^2+0^2)^{1/2}}{5}=0.19$$

Figure 10.4 An example of generational distance (GD) computation.

$$\text{S}=\sqrt{\frac{1}{|Q|}\sum_{i=1}^{|Q|}\left(d_i-\bar{d}\right)^2} \tag{10.14}$$

where $\bar{d}=\sum_{i=1}^{|Q|} d_i/|Q|$ and $d_i=\min_{k\in Q\wedge k\neq i}\sum_{m=1}^{M}\left|f_m^{(i)}-f_m^{(k)}\right|$, which is the minimum value of the sum of the absolute differences in objective function values of the i-th solution and any other solution k in Q. A smaller value of S indicates a more uniform distribution of Q. Figure 10.5 shows one example of computing the spacing measure. For example, in the case of solution A, $d_A=\min((1.6+2.7),(2.8+5.0),(5.8+5.6),(7.2+6.6))=4.3$.

The complexity of computing spacing is $O(|Q|^2)$. However, half of the calculations can be avoided by exploiting the symmetry in distance measures. Also, normalization of the objectives before calculating spacing is essential.

The third measure is called hyper-volume (HV), which is a composite type of measure that evaluates closeness and diversity. It calculates the volume covered by the solutions of the set Q. For each solution $i\in Q$ a hypercube ν_i is constructed with a reference point W and a diagonal corner i. The union of all hypercubes is the HV and it can be calculated as

$$\text{HV}=\text{volume}\left(\cup_{i=1}^{|Q|}\nu_i\right) \tag{10.15}$$

The reference point W can be regarded as the vector of worst objective function values. Algorithms providing solutions that give large values of HV are desirable. Figure 10.6 illustrates the computation of HV through an example.

Solution	f_1	f_2
A	1.2	7.8
B	2.8	5.1
C	4.0	2.8
D	7.0	2.2
E	8.4	1.2

Solution	Distance d_i
A	4.3
B	3.5
C	3.5
D	2.4
E	2.4

$$\bar{d} = 3.22 \text{ and } S = \sqrt{\frac{1}{5}\sum_{i=1}^{5}(d_i - \bar{d})^2} = 0.73$$

Figure 10.5 An example of computing spacing.

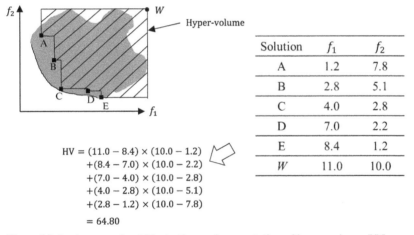

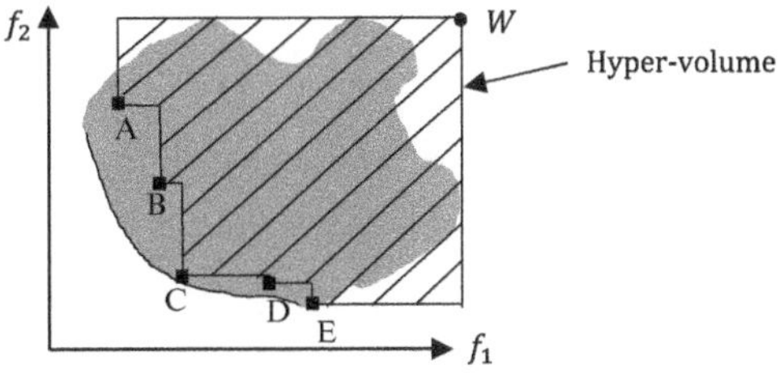

Solution	f_1	f_2
A	1.2	7.8
B	2.8	5.1
C	4.0	2.8
D	7.0	2.2
E	8.4	1.2
W	11.0	10.0

$$\begin{aligned} \text{HV} = {} & (11.0 - 8.4) \times (10.0 - 1.2) \\ & + (8.4 - 7.0) \times (10.0 - 2.2) \\ & + (7.0 - 4.0) \times (10.0 - 2.8) \\ & + (4.0 - 2.8) \times (10.0 - 5.1) \\ & + (2.8 - 1.2) \times (10.0 - 7.8) \\ = {} & 64.80 \end{aligned}$$

Figure 10.6 An example of illustration and computation of hyper-volume (HV).

In case the objective functions values are in different orders of magnitude, for example, if f_1 is an order of magnitude larger than f_2, reducing HV by a unit improvement in f_1 will be much greater than doing that by a unit improvement in f_2. In these cases, one of the following two remedies should be considered: normalize all objective values and use the metric HVR, which is the ratio of the HVs of Q and P^*, $\text{HVR} = \dfrac{\text{HV}(Q)}{\text{HV}(P^*)}$.

10.5 Selection of Preferred Solutions

Once the Pareto-optimal solution set is obtained, higher-level decision making is necessary to choose one or more preferred solutions according to different application backgrounds and specific preferences. The methods for selecting the best compromise solution are called post-optimal techniques, and the methods for selecting a preferred Pareto-optimal region are called optimization-level techniques [16]. In this section, we introduce two popular methods for selecting the best compromise solution. For detailed information about other methods, please refer to [16].

10.5.1 "Min-max" Method

This is a widely used approach for defining a single best-compromise solution. Let z_m^{nad} denote the maximum value of the m-th objective function on set Q. The relative deviation of each objective for each solution is calculated as $r_m = \left(z_m^{nad} - f_m^{(i)}\right) / z_m^{nad}$; then $r_z = \min_m \{r_m\}$ is taken as the representative value of each solution. The solution $\boldsymbol{z}^*$ with the maximum r_z is selected to be the best compromise solution. In practice, DMs should adopt this method when they desire a solution that is representative of the 'center' of the Pareto-front. Figure 10.7 illustrates one example of using the min-max method to choose the best-compromise solution from a bi-objective minimization problem.

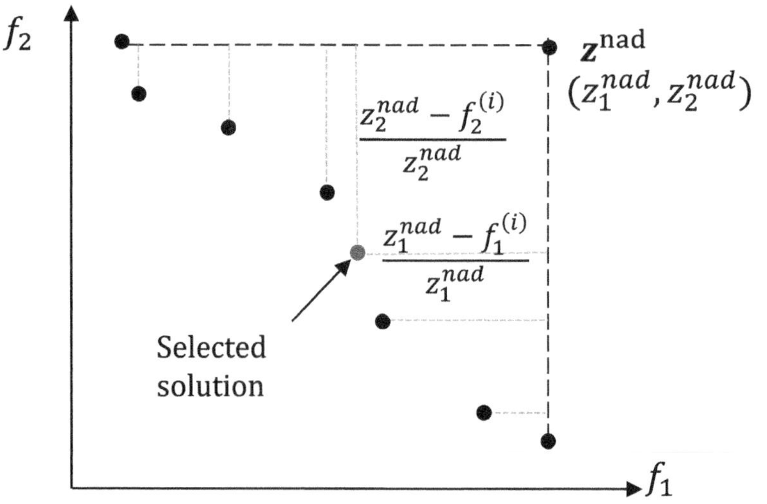

Figure 10.7 Best-compromise solution $\boldsymbol{z}^*$ selected from the Pareto-front by the min-max method, for a two-objective minimization problem.

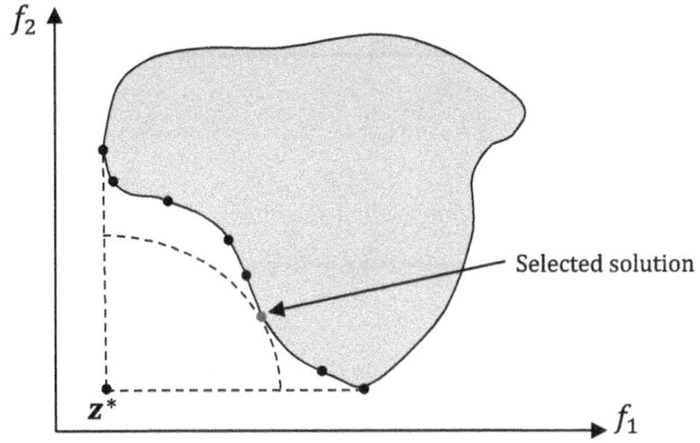

Figure 10.8 Illustration of the compromise Min-Max programming approach, for a two-objective minimization problem with $p=2$.

10.5.2 Compromise Programming Approach

This approach selects the best-compromise solution, which is located closest to a given reference point z. The ideal objective vector z^* can be regarded as the reference point. The following metric is used to measure the distance of the solutions of set Q to the referent point z:

$$d(i,z)=\left(\sum_{m=1}^{M}\left|f_m^{(i)}-f_m^{(z)}\right|^p\right)^{1/p} \tag{10.16}$$

Then the problem of selecting the solutions is converted to the minimization of the distance metric in Equation (10.16). Figure 10.8 illustrates this idea in the two-objective case with $p=2$.

10.6 Guidelines for Solving RAMS+C Optimization Problems

To summarize this chapter, we have drawn out the complete framework to deal with RAMS+C optimization problems, which includes problem formulation, solution method selection and preferred solution(s) selection. Implementation guidelines are presented as the following steps, together with the main points of attention:

1) *Formulate the RAMS+C optimization problem.* As stated in the previous sections, RAMS+C optimization is essentially multi-objective; the following aspects have to be taken into account in the problem formulation:
   a) All objectives need to be analyzed first to reveal the relations between them. For example, the generic unavailability $U(\boldsymbol{x})=1-A(\boldsymbol{x})$ might contain maintainability $M(\boldsymbol{x})$ to describe the unavailability due to test and maintenance activities. The

MOO methods are worth applying only when conflicts exist between at least one pair of objectives; otherwise, a single-objective method can be used to find one solution that optimizes all objectives.

b) DM's preferences for each objective need to be solicited. The preferences can be represented by weights or converted into utility functions. If there is sufficient information about the preferences, in the next step *a priori* solution methods need to be selected; otherwise, the *a posteriori* solution methods have to be chosen. In addition, if the DMs need to intervene during the optimization process, the interactive methods could be considered.

2) *Select appropriate optimization methods.* The choice of the optimization method depends on the formulation of the problem and can largely impact the optimization results. The following aspects need to be considered in this step:

   a) If each objective with the constraints can be analytically solved by a single-objective mathematical programming method in polynomial time, then it is recommended to combine the mathematical programming and the classical MOO problem solution approaches, e.g. weighted-sum (*a priori*) and ε-constraint (*posteriori*) approaches. If the DM provides preference weights (and targets), then weighted-sum (or goal programming) can be used; if the DM wishes to obtain a complete Pareto-optimal-front, then the ε-constraint needs to be used. It is not recommended to use weighted-sum or goal programming to produce the complete Pareto-optimal front due to their disadvantages presented in Sections 10.2.1 and 10.2.2.
   b) If one of the objectives exhibits difficult characteristics, e.g. non-linear, non-convex, NP-hard, then it is recommended to use MOEAs or the classical MOO problem solution approaches combined with single-objective EAs (SOEAs). The original problem can also be relaxed and solved approximately by mathematical programming techniques. This approach is recommended if the practitioner has good knowledge about advanced mathematical optimization theory.
   c) Test more than one method especially for difficult problems, because no one method is the best for all cases, and each method has its own advantages and drawbacks.

3) *Solve the optimization problem.* Depending on the optimization methods selected, the following aspects need to be considered:

   a) In case the exact solutions to all individual objectives can be found via mathematical programming in polynomial time, then the whole problem will be solved once using a priori algorithm and one exact solution will be obtained. If *posteriori* methods are used, then the whole problem needs to be solved multiple times, each under a different setting of the parameters, e.g. ε, and a set of the solutions on the Pareto-optimal front will be obtained.
   b) If MOEAs (or ε-constraint + SOEAs) are used, then multiple simulation runs will be required due to the stochastic nature of these algorithms. The parameters of EAs need to be tuned and the convergence of the EAs ensured. Typically, each simulation run delivers one approximate Pareto-front. To obtain the best results across all runs, all fronts need to be combined, and a final 'front of fronts' will be selected from them. For ε-constraint + SOEAs, each simulation run delivers one

single solution and different ε values need to be explored to obtain the fronts. The final front can be selected from all results.

c) In case that *a priori* MOO problem solution methods are used together with SOEAs, multiple simulation runs, parameter tuning and convergence insurance are also necessary.
d) The performance measures are used to evaluate the quality of the obtained results. If the results are fronts, then the measures presented in Section 10.4 will be useful whereas if the results are single solution points, and statistics such as min, mean, standard deviations, etc. will be used.
e) The optimization results are recommended to be presented, compared, analyzed and validated at this step.

4) *Select the preferred solution(s).* This step is necessary when the *a posteriori* methods are used. First, it is recommended to select the best front from all the final fronts of different methods. The DMs are, therefore, solicited to determine the preferred solution(s) or the method that selects the preferred solution(s).
5) *Validate the results and the procedures.* All results and procedures need to be thoroughly checked to ensure the correctness of the implementation and the meaningfulness and usefulness of the results.

10.7 Exercises

1) Which among the following statements is NOT the difference of MOO from SOO?
   a) MOO has three optimization goals.
   b) MOO also deals with objective space.
   c) MOO tends to obtain a diverse set of optimal solutions.
   d) MOO has artificial fix-ups.
2) For an optimization problem with M objectives, what is the computational complexity of a continuously updated approach for identifying the non-dominated solution set in a given set of size N?
3) What are the main advantages and disadvantages of weighted sum-method and ε-constraint method, respectively?
4) In the following non-dominated solutions for a minimization problem, which is the preferred solution selected by min-max approach?

Solution	f_1	f_2
1	1.1	5.0
2	2.5	3.2
3	3.6	1.3
4	0.0	6.1
5	4.8	0.0

a) Solution 1
b) Solution 2
c) Solution 3
d) Solution 4
e) Solution 5

5) For a two-objective minimization problem, consider the parent $\boldsymbol{P}_t$ and the archive $\boldsymbol{E}_t$ at the t-th generation as follows: The size of $\boldsymbol{P}_t$ and $\boldsymbol{E}_t$ are $N = 4$ and $N_E = 3$, respectively. For the density estimation in fitness assignment, set $k = \sqrt{N + N_E} = 2$. Calculate E_{t+1} by using SPEA2. During the calculation, keep the accuracy to 0.01.

Parent population P_t			Archive E_t		
Solution	Q	f_2	Solution	f_1	f_2
1	3.0	3.5	a	5.0	2.0
2	3.0	4.0	b	4.0	3.5
3	1.0	3.8	c	7.0	3.0
4	6.0	0.0			

References

1 Frank, M.V. (1995). Choosing among safety improvement strategies: A discussion with example of risk assessment and multi-criteria decision approaches for NASA. *Reliability Engineering & System Safety* 49 (3): 311–324.

2 Marseguerra, M. and Zio, E. (2006). Basics of genetic algorithms optimization for RAMS applications. *Reliability Engineering & System Safety* 91 (9): 977–991.

3 Sawaragy, Y., Nakayama, H., and Tanino, T. (1985). *Theory of Multiobjective Optimization*. Orlando, FL: Academic Press.

4 Keeney, R.L. and Raiffa, H. (1993). *Decisions with Multiple Objectives: Preferences and Value Trade-offs*. London, England: Cambridge University Press.

5 Haimes, Y.Y., Lasdon, L.S., and Wismer, D.A. (1971). On a bicriterion formulation of the problems of integrated system identification and system optimization. *IEEE Transactions on Systems Man and Cybernetics* 1: 296–297.

6 Isermann, H. and Steuer, R.E. (1988). Computational experience concerning payoff tables and minimum criterion values over the efficient set. *European Journal of Operational Research* 33 (1): 91–97.

7 Charnes, A., Cooper, W.W., and Ferguson, R.O. (1955). Optimal estimation of executive compensation by linear programming. *Management Science* 1 (2): 138–151.

8 Holand, J.H. (1975). *Adaptation in Natural and Artificial Systems*. Ann Arbor, MI: University of Michigan Press.

9 Deb, K., Pratap, A., Agarwal, S., and Meyarivan, T. (2002). A fast and elitist multiobjective genetic algorithm: NSGA-II. *IEEE Transactions on Evolutionary Computation* 6 (2): 182–197.

10 Zitzler, E. and Thiele, L. (1998). An evolutionary algorithm for multi-objective optimization: The strength pareto approach. *TIK-report* 43.

11 Zitzler, E., Laumanns, M., and Thiele, L. (2001). SPEA2: improving the strength Pareto evolutionary algorithm.

12. Schaffer, J.D. (1985). Multiple objective optimization with vector evaluated genetic algorithms. In: *Proceedings of the 1st international Conference on Genetic Algorithms*, 93–100. L. Erlbaum Associates Inc.

13 Horn, J., Nafpliotis, N., and Goldberg, D.E. (1994). A niched Pareto genetic algorithm for multi-objective optimization. In: *Evolutionary Computation, 1994. IEEE World Congress on Computational Intelligence., Proceedings of the First IEEE Conference on*, 82–87: IEEE.

14 Konak, A., Coit, D.W., and Smith, A.E. (2006). Multi-objective optimization using genetic algorithms: A tutorial. *Reliability Engineering & System Safety* 91 (9): 992–1007.

15 Zitzler, E., Thiele, L., Laumanns, M., Fonseca, C.M., and Da Fonseca, V.G. (2003). Performance assessment of multiobjective optimizers: An analysis and review. *IEEE Transactions on Evolutionary Computation* 7: 117–132.

16 Deb, K. (2001). *Multi-objective Optimization Using Evolutionary Algorithms*. Hoboken, NJ: John Wiley & Sons.

17 van Veldhuizen, D.A. (1999). Multiobjective evolutionary algorithms: Classifications, analyses, and new innovations. Graduate School of Engineering, Air Force Institute of Technology, Air University, Wright-Patterson AFB, OH.

11

Optimization under Uncertainty

Reliability engineering very often deals with the *uncertainties* in the failure and repair processes of components and systems. As presented in Chapter 6, different types of uncertainties exist, which need to be considered in most RAM and Safety criteria (RAMS) optimization. As a result, reliability engineering often adopts and calibrates the methods developed in other domains, e.g. operations research, for solving reliability-related and risk-related optimization problems under uncertainty.

11.1 Stochastic Programming (SP)

Stochastic programming (SP) has been used in a wide variety of application, such as finance planning, power system capacity expansion, airline management planning, location and distribution, and production planning [1]. Its approach is similar to the classical mathematical (or deterministic) programming for solving optimization problems with "random" parameters. Random parameters characterize many real-world applications; in our case of interest, for example, power system operation costs depend on electricity market prices and weather conditions (for renewable generation), which are random. The failure rates of the system's components depend on their operating environments, which are randomly changing, etc. As discussed in Chapter 6, this type of uncertain parameters can be represented by means of random variables, with probability distributions whose parameters are estimated from data.

To include such randomness in the optimization problem, one natural way is to take the expectation of the optimal solutions corresponding to the realizations of the random parameters, as in this example.

Example 11.1 Maintenance manager Ms. Wang is responsible for maintaining a group of 100 water pumps. She needs to decide the maintenance actions to be performed in the next week, with lubrication or repair. For simplicity, assume the condition of all pumps is good or defective. If a pump is good, lubrication is most economical to perform; if it is defective, repair is the most economical action. Table 11.1 shows the cost ($) of the maintenance actions corresponding to each condition.

Assume that the probability of being defective is p for each pump and suppose that $p = 0.1$. Because lubrication is optimal for good pumps and repair is optimal for defective ones, is it optimal to lubricate 90% of the pumps and repair 10% of them? If yes, then in this plan, the expected total cost is $(10 + 600 \times 0.1) \times 0.9 + (50 + 100 \times 0.1) \times 0.1 = 69$ dollars.

Let us check if the solution above is correct. Given that the expected costs of performing different actions are $10 + 600 \times 0.1 = 70\$$ for lubrication and $50 + 100 \times 0.1 = 60\$$ for repair, the formulation of this optimization problem is the following:

$$\max\ 70x_1 + 60x_2$$

$$\text{s.t. } x_1 + x_2 = 1$$

$$x_1,\ x_2 \geq 0$$

Obviously, the solution is doing repair for ALL, and the expected cost of this plan is 60 dollars against the previous one of 69 dollars.

From this example, we can see that the optimal solution in general is not equal to the "average" of the best decisions for each specific future outcome. The correct way is to optimize the expectation of the objective values taking into account the random parameters. SP is the way to deal with this. With SP, randomness is represented in terms of the random experiments with outcomes denoted by ω. The set of all outcomes is represented by Ω. Outcomes may be combined into subsets of Ω, which are called events. Let $\mathcal{B}$ denote a collection of all random events. For each event $B \in \mathcal{B}$, a value $\Pr(B)$ is associated, called its probability.

Under this setting, the objective (or constraint) function becomes the expectation of the random objective function (or constraint):

$$\boldsymbol{F}(\boldsymbol{x}) = \boldsymbol{E}(\boldsymbol{f}(\boldsymbol{x}, \xi(\omega))) \tag{11.1}$$

or the probability of the event $B(\boldsymbol{x})$:

$$\boldsymbol{F}(\boldsymbol{x}) = \mathbf{Pr}(\xi(\omega) \in \boldsymbol{B}(\boldsymbol{x})) \tag{11.2}$$

Table 11.1 The cost ($) of different maintenance actions under each condition.

Action **Condition**	**Lubrication**	**Repair**
Good	10	50
Defect	600*	100

*It includes the expected loss incurred by uncorrected defects

where $\boldsymbol{x} = (x_1, x_2, \ldots, x_n)$ is a vector of decision variables and $\boldsymbol{\xi} = (\xi_1, \xi_2, \ldots, \xi_m)$ is a vector of random parameters. The second formulation in Equation (11.2) is also referred to as chance-constrained problem.

In the following of this section, we introduce three representative types of SP, namely: two-stage stochastic linear programs with fixed recourse, multi-stage stochastic programs with recourse, and probabilistic or chance-constrained programs.

11.1.1 Two-stage Stochastic Linear Programs with Fixed Recourse

The most widely applied and studied SP models are two-stage linear programs. The decision maker takes certain actions in the first stage after which a random event occurs affecting the outcome of the first-stage decision. A *recourse* decision can be made in the second stage to compensate for any bad effects that might occur as a result of the first-stage decision, which will forbid the random event occurrence.

The formulation of classical two-stage stochastic linear programs with fixed recourse [2,3] is as follows:

$$\min z = \boldsymbol{c}^T \boldsymbol{x} + \mathrm{E}_\xi \left[\min \boldsymbol{q}(\omega)^T \mathbf{y}(\omega) \right] \tag{11.3a}$$

$$\text{s.t. } \boldsymbol{Ax} = \boldsymbol{b} \tag{11.3b}$$

$$\boldsymbol{T}(\omega)\boldsymbol{x} + \boldsymbol{W}\mathbf{y}(\omega) = \boldsymbol{h}(\omega) \tag{11.3c}$$

$$\boldsymbol{x} \geq 0,\ \boldsymbol{y}(\omega) \geq 0 \tag{11.3d}$$

where $\boldsymbol{x}$ is a $n_1 \times 1$ vector of decisions to be taken without full information on the subsequent random events. These decisions are called first-stage decisions. Corresponding to $\boldsymbol{x}$ are the first-stage vectors and matrix $\boldsymbol{c}, \boldsymbol{b}$ and $\boldsymbol{A}$, of sizes $n_1 \times 1$, $m_1 \times 1$, and $m_1 \times n_1$, respectively. In the second stage, a number of random events $\omega \in \Omega$ may be realized. For a given realization ω, the second-stage problem parameters $\boldsymbol{q}(\omega)$, $\boldsymbol{h}(\omega)$ and $\boldsymbol{T}(\omega)$ become known where $\boldsymbol{q}(\omega)$ is $n_2 \times 1$, $\boldsymbol{h}(\omega)$ is $m_2 \times 1$, and $\boldsymbol{T}(\omega)$ is $m_2 \times n_1$. The *recourse matrix* $\boldsymbol{W}$ is the known matrix of size $m_2 \times n_2$, which is assumed to be fixed. Then second-stage or corrective actions $\boldsymbol{y}(\omega)$ are taken. The ξ is the vector formed by components of $\boldsymbol{q}^T$, $\boldsymbol{h}^T$ and $\boldsymbol{T}$. The notation E_ξ denotes the mathematical expectation with respect to $\boldsymbol{\xi}$.

The objective function Equation (11.3a) contains a deterministic term $\boldsymbol{c}^T \boldsymbol{x}$ and the expectation of the second-stage objective $\boldsymbol{q}(\omega)^T \mathbf{y}(\omega)$ taken over all realizations of the random event ω. This second-stage term is the more difficult one because for each ω, the value $\boldsymbol{y}(\omega)$ is the solution of a linear program. To address this issue, a deterministic equivalent (DE) program is developed in the following.

For a given realization ω, let

$$Q(\boldsymbol{x}, \boldsymbol{\xi}(\omega)) = \min_{\boldsymbol{y}(\omega)} \{ \boldsymbol{q}(\omega)^T \mathbf{y}(\omega) \mid \boldsymbol{Wy}(\omega) = \boldsymbol{h}(\omega) - \boldsymbol{T}(\omega)\boldsymbol{x}, \boldsymbol{y}(\omega) \geq 0 \} \tag{11.4}$$

be the second-stage value function, i.e. *recourse function*. The expected second-stage value function, i.e. *expected recourse function*, is defined as

$$\mathcal{Q}(\boldsymbol{x}) = \mathrm{E}_\xi Q(\boldsymbol{x}, \boldsymbol{\xi}(\omega)) \tag{11.5}$$

Then we can rewrite the problem in Equation (11.3) in terms of only $\mathbf{x}$ as follows

$$\min z = \boldsymbol{c}^T \boldsymbol{x} + \mathcal{Q}(\boldsymbol{x}) \tag{11.6a}$$

$$\text{s.t. } \boldsymbol{A}\boldsymbol{x} = \boldsymbol{b} \tag{11.6b}$$

$$\boldsymbol{x} \geq 0 \tag{11.6c}$$

This representation is named the DE of the original stochastic program. For a given realization ω, it is a non-linear program due to the 'min' operation in the recourse function.

To solve the problem in Equation (11.3), the most difficult part is the evaluation of the expected recourse function $\mathcal{Q}(\boldsymbol{x})$ because it often needs a large number of realizations of the random parameters ξ. To deal with this problem, the key idea is to approximate $\mathcal{Q}(\boldsymbol{x})$ using different approaches, e.g. sampling and decomposition. In the following, we introduce the solution techniques based upon the above two approaches.

Sample Average Approximation

In theory, we would want to obtain a solution with reasonable accuracy and acceptable solving time. A possible way to unite these two conflicting goals is by randomization, i.e., Monte Carlo sampling techniques. Suppose the total number of the possible realizations of random parameters $\boldsymbol{\xi}$ is large or infinite, and we can generate random samples $\boldsymbol{\xi}^1,\dots, \boldsymbol{\xi}^N$ of the random vector $\boldsymbol{\xi}$. Given these samples, we can approximate the expectation function $\mathcal{Q}(x) = \mathrm{E}_\xi Q(\boldsymbol{x}, \boldsymbol{\xi}(\omega))$ by the average

$$\hat{\mathcal{Q}}_N(\boldsymbol{x}) = N^{-1} \sum_{j=1}^{N} \boldsymbol{Q}(\boldsymbol{x}, \xi^j) \tag{11.7}$$

and, thus, the problem in Equation (11.3) can be rewritten as

$$\min_{x \in X} \hat{g}_N(\boldsymbol{x}) = \boldsymbol{c}^T \boldsymbol{x} + N^{-1} \sum_{j=1}^{N} \boldsymbol{Q}(\boldsymbol{x}, \xi^j) \tag{11.8}$$

This technique is fundamental and it can be used to solve general stochastic programs, e.g. with non-linear objective functions and constraints. Extensions of this technique include sample average approximation with an L-shaped method [4], the stochastic decomposition method [5,6], and the stochastic quasi-gradient [7].

Example 11.2 The planning horizon is $J = \{0, 1, 2, 3, 4, 5\}$ in arbitrary units of time. A system consists of one component with known failure distribution, i.e. the lifetime equals to 1, 2, 3 with a probability of 0.6, 0.3 and 0.1. We assume this component must be replaced when it fails. Each planned maintenance is economical with the repair cost $d = 2$ in arbitrary unit of cost. If the failure occurs without a maintenance plan, the replacement generates the cost $c = 10$. At the current time 1, the component is working. So, how to select the replacement decision at current time to minimize the total expected cost for the entire planning period J?

The current system state is represented as $(t, \boldsymbol{\xi}, \boldsymbol{a})$ where t is the current time, the component state $\boldsymbol{\xi} = 1$ if component is failed and 0 otherwise, and $\boldsymbol{a}$ is the age of the component. Start the timer at the current time and set the remaining time period as $J = [0, 1, 2, 3, 4]$ with $T = 4$. Given the current state, the expected minimum total maintenance cost $\boldsymbol{f}_t(\boldsymbol{\xi}, \boldsymbol{a})$ at current time is formally obtained by solving:

$$\boldsymbol{f}_t(\boldsymbol{\xi}, \boldsymbol{a}) = \min\ d\boldsymbol{x} + Q_t^a(\boldsymbol{x}) \tag{11.9a}$$

$$\text{s.t. } \boldsymbol{x} \geq \boldsymbol{\xi} \tag{11.9b}$$

$$\boldsymbol{x} \in \{0, 1\} \tag{11.9c}$$

Let $x = 1$ if we decide to replace the component at the current time, and 0 otherwise. The total maintenance cost in Equation (11.9a) of a decision is the sum of the current maintenance cost and the future cost. The expected minimum future cost given the system state is represented as $Q_t^a(\boldsymbol{x})$.

The maximum number of components used in the remaining period is $T + 1 = 5$. So, the possible components are $\mathcal{R} = \{1,2,3,4,5\}$. Given the current age a and the failure distribution, all possible scenarios of this problem are defined as $w \in \Omega$ with probability $p(w)$. Each life of the individual $r \in \mathcal{R}$ in scenario w is T_r^w. Then we can formulate the extensive form with the second-stage variables given by

$$y_{tr}^w = \begin{cases} 1, & \text{if individual } r \text{ is replaced at} \\ & \text{or before } t \text{ in scenario } w \quad \forall t \in J, r \in \mathcal{R}, w \in \Omega \\ 0, & \text{otherwise} \end{cases}$$

The deterministic formulation is

$$\min \sum_{w \in \Omega} p(w) \left(\sum_{r \in \mathcal{R}} c y_{tr}^w + (d - c) x \right)$$

$$\text{s.t. } y_{tr}^w \leq y_{t+1,r}^w,\ t \in J \setminus \{T\},\ r \in \mathcal{R},\ w \in \Omega$$

$$y_{t+1,r+1}^w \leq y_{tr}^w,\ t \in J \setminus \{T\},\ r \in \mathcal{R} \setminus \{q\},\ w \in \Omega$$

$$y_{tr}^w \leq y_{t+T_{r+1}^w, r+1}^w,\ r \in \mathcal{R} \setminus \{q\}, w \in \Omega,\ t \in \{0, \ldots, T - T_{r+1}^w\}$$

$$y_{T_1^w,1}^w = 1,\ w \in \Omega \text{ and if } T_1^w \leq T$$

$$y_{0r}^w = 0,\ r \in \mathcal{R} \setminus \{1\}, w \in \Omega$$

$$x = y_{01}^w,\ w \in \Omega$$

$$x \geq \xi$$

$$y_{tr}^w \in \{0,1\},\ t \in J,\ r \in \mathcal{R},\ w \in \Omega$$

$$x \in \{0,1\}$$

where q denotes the last individual in $\mathcal{R}$.

To avoid solving this large integer programming (IP) for all possible scenarios, we use the sample average approximation method to approximate this problem. We generate $|\Omega| = 15$ random realizations of $T_r^w, \forall r \in \mathcal{R}$. The approximated objective value is 23.333. The decision is 1 indicating that the replacement is implemented at the current time.

L-shaped method

Decomposition methods make use of the special structure of this stochastic program to improve the effectiveness of the solution algorithms. The most common decomposition technique is called L-shaped technique [8]. Other decomposition procedures include inner linearization, Dantzig-Wolfe decomposition, etc. The basic idea of the L-shaped method is to approximate the non-linear term (the recourse function) in the objective by a linear one. Therefore, the master problem of $\mathbf{x}$ is reconstructed as the first-stage problem plus the outer linearization of the recourse function. Thus, the recourse function is only evaluated in the sub-problem to avoid numerous evaluations.

Suppose the random vector $\boldsymbol{\xi}$ has finite possible realizations ξ_k with probability p_k for $k = 1, \ldots, K$. We denote the second-stage decision vector as $\boldsymbol{y}_k$ under each realization of $\xi_k = (q_k, h_k, T_k)$ and $k = 1, \ldots, K$. The recourse function can be rewritten as

$$\mathcal{Q}(x) = E_\xi Q(\boldsymbol{x}, \xi(\omega)) = \sum_{k=1}^{K} p_k Q(\boldsymbol{x}, \xi_k)$$

Then the large-scale DE linear program (i.e., the extensive form) of Equation (11.6) is defined in the following way:

$$\min_{x, y_1, \ldots, y_K} \ z = \boldsymbol{c}^T \boldsymbol{x} + \sum_{k=1}^{K} p_k \boldsymbol{q}_k^T \boldsymbol{y}_k \tag{11.10a}$$

$$\text{s.t. } \boldsymbol{A}\boldsymbol{x} = \boldsymbol{b} \tag{11.10b}$$

$$\boldsymbol{T}_k \boldsymbol{x} + \boldsymbol{W} \boldsymbol{y}_k = \boldsymbol{h}_k, k = 1, \ldots, K \tag{11.10c}$$

$$\boldsymbol{x} \geq 0, \ \boldsymbol{y}_k \geq 0, k = 1, \ldots, K \tag{11.10d}$$

The special structure of the constraint matrix for the two-stage extensive form is shown as the block matrix

$$\begin{pmatrix} \boldsymbol{A} & & & & \\ \boldsymbol{T}_1 & \boldsymbol{W} & & & \\ \boldsymbol{T}_2 & & \boldsymbol{W} & & \\ \vdots & & & \ddots & \\ \boldsymbol{T}_K & & & & \boldsymbol{W} \end{pmatrix}$$

Taking the dual of the extensive form, the constraint matrix is rewritten as a block-angular structure:

$$\begin{pmatrix} A^T & T_1^T & T_2^T & \cdots & T_K^T \\ & W^T & & & \\ & & W^T & & \\ & & & \ddots & \\ & & & & W^T \end{pmatrix}$$

This is a large linear programming (LP) problem and has special structure. Therefore, we can solve this problem by a Benders decomposition [9] of the primal or a Dantzig-Wolfe decomposition [10] of the dual to reduce the computation.

For Equation (11.6), the recourse function $\mathcal{Q}(\mathbf{x})$ is approximated with an artificial variable θ, which represents the lower bound for $\mathcal{Q}(\mathbf{x})$. Now we have

$$\theta \geq \mathcal{Q}(\hat{\boldsymbol{x}}) + \boldsymbol{u}^T \ (\boldsymbol{x} - \hat{\boldsymbol{x}})$$

where $\boldsymbol{u} = -\sum_{k=1}^{K} p_k \boldsymbol{T}_k^T \boldsymbol{\lambda}_k^* \in \partial \mathcal{Q}(\hat{\boldsymbol{x}})$ and $\boldsymbol{\lambda}_k^*$ is the optimal dual solution of the recourse problem in scenario k with $\hat{\boldsymbol{x}}$. So, given a feasible decision $\hat{\boldsymbol{x}}$, we can build up the linear approximation of $\mathcal{Q}(\boldsymbol{x})$ by

$$\theta \geq \hat{e} - \boldsymbol{Ex}$$

where $E = \sum_{k=1}^{K} p_k \cdot \left(\boldsymbol{\lambda}_k^*\right)^T \boldsymbol{T}_k$ and $\hat{e} = \sum_{k=1}^{K} p_k \cdot \left(\boldsymbol{\lambda}_k^*\right)^T \boldsymbol{h}_k$.

(Hints:

1. $\boldsymbol{u} = -\sum_{k=1}^{K} p_k \boldsymbol{T}_k^T \boldsymbol{\lambda}_k^* \in \partial \mathcal{Q}(\boldsymbol{x})$

Because $\mathcal{Q}(x) = E_\xi Q(\boldsymbol{x}, \boldsymbol{\xi}(\omega)) = \sum_{k=1}^{K} p_k Q(\boldsymbol{y}, \xi_k)$, then $\partial \mathcal{Q}(\boldsymbol{x}) = \sum_{k=1}^{K} p_k \partial Q(\boldsymbol{x}, \xi_k)$. We also have that $Q(\boldsymbol{x}, \xi_k) = \min\left\{\boldsymbol{q}_k^T \boldsymbol{y}_k : \boldsymbol{W}\boldsymbol{y}_k = \boldsymbol{h}_k - \boldsymbol{T}_k\boldsymbol{x}, \boldsymbol{y}_k \geq 0 \; k = 1, \ldots, K\right\}$. Consider the dual problem of this problem with dual variable λ_k; then

$$\begin{aligned}&\min\left\{\boldsymbol{q}_k^T \boldsymbol{y}_k : \boldsymbol{W}\boldsymbol{y}_k = \boldsymbol{h}_k - \boldsymbol{T}_k\boldsymbol{x}, \boldsymbol{y}_k \geq 0 \; k = 1, \ldots, K\right\} \\ &\quad = \max\left\{\left(\boldsymbol{h}_k - \boldsymbol{T}_k\boldsymbol{x}\right)^T \boldsymbol{\lambda}_k : \ \boldsymbol{\lambda}_k^T \boldsymbol{W} \leq \boldsymbol{q}_k^T\right\}\end{aligned}$$

So, $Q(\boldsymbol{x}, \xi_k) = \left(\boldsymbol{h}_k - \boldsymbol{T}_k\boldsymbol{x}\right)^T \boldsymbol{\lambda}_k^*$ where $\boldsymbol{\lambda}_k^*$ is the optimal solution of the dual problem. We get that $-\boldsymbol{T}_k^T \boldsymbol{\lambda}_k^* \in \partial Q(\boldsymbol{x}, \xi_k)$. $\boldsymbol{u} = -\sum_{k=1}^{K} p_k \cdot \boldsymbol{T}_k^T \boldsymbol{\lambda}_k^* \in \partial \mathcal{Q}(\boldsymbol{x})$ is obtained. Given the decision $\hat{\mathbf{x}}$, the vector $-\sum_{k=1}^{K} p_k \cdot \boldsymbol{T}_k^T \boldsymbol{\lambda}_k^*$ is one of the directional vectors $\partial \mathcal{Q}(\hat{\boldsymbol{x}})$ where $\boldsymbol{\lambda}_k^*$ is associated with $\hat{\boldsymbol{x}}$.

2. $\theta \geq \hat{e} - \boldsymbol{Ex}$

Given $\theta \geq \mathcal{Q}\left(\hat{\boldsymbol{x}}\right) + u^T \left(\boldsymbol{x} - \hat{\boldsymbol{x}}\right)$ and $\boldsymbol{u} = -\sum_{k=1}^{K} p_k \cdot \boldsymbol{T}_k^T \boldsymbol{\lambda}_k^*$.

$$Q(\hat{x}) + u^T(x - \hat{x}) = \sum_{k=1}^{K} p_k Q(\hat{x}, \xi_k) - \sum_{k=1}^{K} p_k \cdot (T_k^T \lambda_k^*)^T (x - \hat{x})$$
$$= \sum_{k=1}^{K} (h_k - T_k \hat{x})^T \lambda_k^* - \sum_{k=1}^{K} p_k \cdot (T_k^T \lambda_k^*)^T (x - \hat{x})$$
$$= \sum_{k=1}^{K} p_k \cdot (\lambda_k^*)^T h_k - \sum_{k=1}^{K} p_k \cdot (\lambda_k^*)^T T_k x$$

Then the lower bound of recourse function $\theta \geq e - Ex$ is obtained).

To guarantee the decision $\hat{x}$ is feasible for the recourse problem, that is $\hat{x} \in K_2 = \{x \mid Q(x) < \infty\}$, we have to check its feasibility first. Consider the linear program

$$\min\ w_k' = e^T v^+ + e^T v^-$$

$$\text{s.t. } Wy + Iv^+ + Iv^- = h_k - T_k \hat{x}$$

$$y, v^+, v^- \geq 0$$

where $e^T = (1,\ldots,1)$. If $w_k' \leq 0$ for all scenarios k and $\hat{x}$ is feasible for the recourse problem. Otherwise, there exists scenario k and $w_k' > 0$, so that $\hat{x}$ is infeasible.

To cut this infeasible solution $\hat{x}$, we generate the feasible cuts. Consider the dual problem and let σ_k^* represent the dual optimal solution: Therefore, this decision has the property $(\sigma_k^*)^T (h_k - T_k \hat{x}) > 0$ and $(\sigma_k^*)^T W \leq 0$. However, for all $x \in K_2$, there exist $y \geq 0$ subject to $Wy = h_k - T_k x$. So, $(\sigma_k^*)^T (h_k - T_k x) = (\sigma_k^*)^T Wy \leq 0$, and the inequality $(\sigma_k^*)^T (h_k - T_k x) \geq 0$ can cut this infeasible solution $\hat{x}$. The algorithm of L-shaped method [8] is presented as follows.

Standard L-shaped Algorithm

Step 0. Set $r = s = v = 0$.

Step 1. Set $v = v + 1$. Solve the linear program:

$$\min\ z = c^T x + \theta \tag{11.11a}$$

$$\text{s.t. } Ax = b \tag{11.11b}$$

$$D_f x \geq d_f, \quad f = 1,\ldots,r \tag{11.11c}$$

$$E_g x + \theta \geq e_g, \quad g = 1,\ldots,s \tag{11.11d}$$

$$x \geq 0, \quad \theta \in R \tag{11.11e}$$

The optimal solution is (x^v, θ^v). If no constraint exists, set θ^v as $-\infty$; that is, θ^v is not considered in the LP.

Step 2. Add feasibility cuts.
If $\mathbf{x}^v \in K_2$, go to Step 3. Otherwise, add the cut(s) in Equation (11.11d) and return to Step 1.

For $k = 1,\ldots,K$ solve the linear program

$$\min\ w' = \boldsymbol{e}^T \boldsymbol{v}^+ + \boldsymbol{e}^T \boldsymbol{v}^-$$

$$\text{s.t. } \boldsymbol{Wy} + \boldsymbol{Iv}^+ - \boldsymbol{Iv}^- = \boldsymbol{h}_k - \boldsymbol{T}_k \boldsymbol{x}^\nu$$

$$\boldsymbol{y},\ \boldsymbol{v}^+, \boldsymbol{v}^- \geq 0$$

where $\boldsymbol{e}^T = (1,\ldots,1)$. If there exists k, the associated $w' > 0$. Then the constraint in Equation (11.11c) is generated with $\boldsymbol{D}_{f+1} = \left(\sigma_k^\nu\right)^T \boldsymbol{T}_k$ and $d_{f+1} = \left(\sigma_k^\nu\right)^T \boldsymbol{h}_k$ where σ_k^ν contains the associated dual multipliers. Set $f = f+1$, add the constraint to Equation (11.11) and return to Step 1. Otherwise, go to Step 3.

Step 3. Add optimality cuts.
For $k = 1,\ldots,K$, solve the sub-problem

$$\min\ w = \boldsymbol{q}_k^T \boldsymbol{y}$$

$$\text{s.t. } \boldsymbol{Wy} = \boldsymbol{h}_k - \boldsymbol{T}_k \boldsymbol{x}^\nu$$

$$\boldsymbol{y} \geq 0$$

Let $\boldsymbol{\lambda}_k^\nu$ be the optimal dual multipliers of the sub-problem given k and x^ν.

Let $w^\nu = \sum_{k=1}^{K} p_k \cdot \left(\boldsymbol{\lambda}_k^\nu\right)^T \left(\boldsymbol{h}_k - \boldsymbol{T}_k \boldsymbol{x}^\nu\right)$. If $\theta^\nu \geq w^\nu$, stop and $\boldsymbol{x}^\nu$ becomes the optimal solution. Otherwise, add the constraint to Equation (11.11d) with $\boldsymbol{E}_{g+1} = \sum_{k=1}^{K} p_k \cdot \left(\boldsymbol{\lambda}_k^\nu\right)^T \boldsymbol{T}_k$ and $e_{g+1} = \sum_{k=1}^{K} p_k \cdot \left(\boldsymbol{\lambda}_k^\nu\right)^T \boldsymbol{h}_k$ into the problem (11.11). Set $g = g+1$, and return to Step 1.

The L-shaped method has three main steps: (1) the master problem in Equation (11.11a) determines the first-stage decision $\mathbf{x}^\nu$ of the deterministic part of the objective and is sent to the second stage; (2) feasibility cuts in Equation (11.11c) are generated based on the second-stage feasibility; (3) optimality cuts in Equation (11.11d) are generated to give the linear approximations to the expected recourse function $\mathcal{Q}(\boldsymbol{x})$.

Example 11.3 There are three types of components available in the market, which can be used for the system. The costs of the three components are 0.4, 0.8, and 0.6 per arbitrary unit price, respectively. After the system works for a period of time t, the proportion of the three types that has not failed is $a = (a_1, a_2, a_3)$. Assume that the survival rate a is fixed under the same environment condition and changes as the conditions change. The system requires that at least 90% of components should be working at time t. If the requirement is not satisfied, a maintenance plan is used to guarantee this system requirement. The maintenance costs are only related to the environment conditions. The decision makers (DMs) wants to decide which percentage of these three types of components to buy, so they can minimize the purchase cost and the expected maintenance cost.

We denote the random condition as $\boldsymbol{\xi}$ and the realization as $\boldsymbol{\xi}_k, \forall k \in K$ with probability p_k. The survival rate and the maintenance cost take on the values (0.7, 0.6, 0.5) and 5 with probability 0.1, (0.7, 0.6, 0.7) and 3 with probability 0.4, (0.5, 0.7, 0.8) and 2 with probability 0.2, and (0.6, 0.7, 0.9) and 4 with probability 0.3. The extensive form becomes

$$\min 0.4x_1 + 0.8x_2 + 0.6x_3 + E_\xi(q\boldsymbol{y})$$
$$\boldsymbol{T}^T\boldsymbol{x} + \boldsymbol{y} \geq \boldsymbol{h}$$
$$\boldsymbol{e}^T\boldsymbol{x} = 1$$
$$\boldsymbol{x}, \boldsymbol{y} \geq 0$$

where $h_k = 0.9$, $\boldsymbol{T}^T(\xi)$ and $\boldsymbol{q}(\xi)$ denote the survival rate and the maintenance cost, respectively, and $e = (1,1,1)^T$.

Use the L-shaped method to solve this problem. In this example, the second stage is always satisfied because $\boldsymbol{h} - \boldsymbol{T}^T\boldsymbol{x} \leq 1$ and $\boldsymbol{y} \geq \boldsymbol{h} - \boldsymbol{T}^T\boldsymbol{x}$ always exist. Step 2 can be omitted.

Iteration 1:
Step 1. Ignoring θ, the master program is $\min\{0.4x_1 + 0.8x_2 + 0.6x_3 \mid x_1 + x_2 + x_3 = 1,\ x_1 \geq 0, x_2 \geq 0, x_3 \geq 0\}$. The solution is $\boldsymbol{x}^1 = (1,0,0)^T$ and $\theta^1 = -\infty$.

Step 3.

- For $\xi = \xi_1$, solve the sub-problem

$$w_1 = \min\{5y \mid 0.7 + y \geq 0.9, y \geq 0\}$$

  The solution is $y = 0.2, \lambda_1 = 5$.
- For $\xi = \xi_2$, solve the sub-problem

$$w_2 = \min\{3y \mid 0.7 + y \geq 0.9, y \geq 0\}$$

  The solution is $y = 0.2$, $\lambda_2 = 3$.
- For $\xi = \xi_3$, solve the sub-problem

$$w_3 = \min\{2y \mid 0.5 + y \geq 0.9, y \geq 0\}$$

  The solution is $y = 0.4$, $\lambda_3 = 2$.
- For $\xi = \xi_4$, solve the sub-problem

$$w_4 = \min\{4y \mid 0.6 + y \geq 0.9, y \geq 0\}$$

  The solution is $y = 0.3$, $\lambda_4 = 4$.

Using $h_k = 0.9, \forall k \in K$, we get that

$$\boldsymbol{E}_1 = 0.1\lambda_1\boldsymbol{T}_1 + 0.4\lambda_2\boldsymbol{T}_2 + 0.2\lambda_3\boldsymbol{T}_3 + 0.3\lambda_4\boldsymbol{T}_4 = (2.11, 2.14, 2.49)$$
$$e_1 = 0.1\lambda_1 h_1 + 0.4\lambda_2 h_2 + 0.2\lambda_3 h_3 + 0.3\lambda_4 h_4 = 2.97$$
$$w^1 = e_1 - \boldsymbol{E}_1\boldsymbol{x}^1 = 0.86 > \theta^1$$

Finally, as $w^1 > \theta^1$, add the cut

$$2.11x_1 + 2.14x_2 + 2.49x_3 + \theta^1 \geq 2.97.$$

Iteration 2:
Step 1. Solve the master program

$$\min\{0.4x_1 + 0.8x_2 + 0.6x_3 + \theta \mid x_1 + x_2 + x_3 = 1, 2.11x_1 + 2.14x_2 + 2.49x_3 + \theta \geq 2.97,\ x_1 \geq 0, x_2 \geq 0, x_3 \geq 0\}$$

The solution is $x^2 = (0,0,1)^T$ and $\theta^2 = 0.48$.

Step 3.

- For $\xi = \xi_1$, solve the sub-problem

$$w_1 = \min\{5y \mid 0.5x_3 + y \geq 0.9, y \geq 0\}$$

  The solution is $y = 0.4$, $\lambda_1 = 5$.
- For $\xi = \xi_2$, solve the sub-problem

$$w_2 = \min\{3y \mid 0.7 + y \geq 0.9, y \geq 0\}$$

  The solution is $y = 0.2$, $\lambda_2 = 3$.
- For $\xi = \xi_3$, solve the sub-problem

$$w_3 = \min\{2y \mid 0.8 + y \geq 0.9, y \geq 0\}$$

  The solution is $y = 0.1$, $\lambda_3 = 2$.
- For $\xi = \xi_4$, solve the sub-problem

$$w_4 = \min\{4y \mid 0.9 + y \geq 0.9, y \geq 0\}$$

  The solution is $y = 0$, $\lambda_4 = 0$.

Using $h_k = 0.9 \forall k \in K$, we get that

$$\boldsymbol{E}_2 = 0.1\lambda_1\boldsymbol{T}_1 + 0.4\lambda_2\boldsymbol{T}_2 + 0.2\lambda_3\boldsymbol{T}_3 + 0.3\lambda_4\boldsymbol{T}_4 = (1.39, 1.3, 1.41)$$

$$e_2 = 0.1\lambda_1 h_1 + 0.4\lambda_2 h_2 + 0.2\lambda_3 h_3 + 0.3\lambda_4 h_4 = 1.89$$

$$w^2 = e_2 - \boldsymbol{E}_2\boldsymbol{x}^2 = 0.48 = \theta^2$$

Stop.

The outcome is the following: $x^2 = (0,0,1)^T$ is the optimal solution and the optimal objective value is 1.08.

11.1.2 Multi-stage Stochastic Programs with Recourse

The previous sections focused on stochastic programs with two stages. However, most practical decision problems involve a sequence of decisions that react to outcomes that evolve over time. In this section, the SP approach to multi-stage problems [11] is presented. The linear, fixed recourse, finite horizon framework is used due to its widespread implementation [12]. Its formulation is presented as follows (the transposes indexes are suppressed in the notation when they are clear from the context to avoid excessive notation):

$$\min z = \boldsymbol{c}^1\boldsymbol{x}^1 + \mathrm{E}_{\xi^2}\left[\min \boldsymbol{c}^2(\omega)\boldsymbol{x}^2(\omega^2) + \ldots + \mathrm{E}_{\xi^H}\left[\min \boldsymbol{c}^H(\omega)\mathbf{x}^H(\omega^H)\right]\ldots\right] \tag{11.12}$$

$$\text{s.t.} \quad \boldsymbol{W}^1\boldsymbol{x}^1 = \boldsymbol{h}^1$$

$$\boldsymbol{T}^1\left(\omega^2\right)\boldsymbol{x}^1 + \boldsymbol{W}^2\boldsymbol{x}^2\left(\omega^2\right) = \boldsymbol{h}^2\left(\omega\right)$$

$$\ldots$$

$$\boldsymbol{T}^{H-1}\left(\omega^H\right)\boldsymbol{x}^{H-1}\left(\omega^{H-1}\right) + \boldsymbol{W}^H\boldsymbol{x}^H\left(\omega^H\right) = \boldsymbol{h}^H\left(\omega\right)$$

$$\boldsymbol{x}^1 \geq 0; \quad \boldsymbol{x}^t\left(\omega^t\right) \geq 0, \quad t = 2,\ldots,H$$

where $\boldsymbol{c}^1$ is a known vector in $\Re^{n_1}$, $\boldsymbol{h}^1$ is a known vector in $\Re^{m_1}$, $\xi^t\left(\omega\right)^T$ is the vector formed by components of $\boldsymbol{c}^t\left(\omega\right)^T, \boldsymbol{h}^t\left(\omega\right)^T$ and $\boldsymbol{T}^{t-1}\left(\omega^t\right)$, and each $\boldsymbol{W}^t$ is a known $m_t \times n_t$ matrix. The decisions $\boldsymbol{x}$ depend on the history up to time t, which is indicated by ω^t.

The DE form of this problem can be described in terms of dynamic programming (DP) [1]. If the stages are 1 to H, we can define states as $\boldsymbol{x}^t\left(\omega^t\right)$. For the terminal conditions, we have:

$$Q^H\left(\boldsymbol{x}^{H-1}, \xi^H\left(\omega\right)\right) = \min \boldsymbol{c}^H\left(\omega\right)\boldsymbol{x}^H\left(\omega\right) \tag{11.13a}$$

$$\text{s.t. } \boldsymbol{W}^H\boldsymbol{x}^H\left(\omega\right) = \boldsymbol{h}^H\left(\omega\right) - \boldsymbol{T}^{H-1}\left(\omega\right)\boldsymbol{x}^{H-1} \tag{11.13b}$$

$$\boldsymbol{x}^H\left(\omega\right) \geq 0 \tag{11.13c}$$

Solutions for other stages can be obtained with a backward recursion, letting - $\mathcal{Q}^{t+1}\left(\boldsymbol{x}^t\right) = \mathrm{E}_{\xi^{t+1}}\left[Q^{t+1}\left(\boldsymbol{x}^t, \xi^{t+1}\left(\omega\right)\right)\right]$ for all t to obtain the recursion for $t = 2,\ldots,H-1$

$$Q^t\left(\boldsymbol{x}^{t-1}, \xi^t\left(\omega\right)\right) = \min \boldsymbol{c}^t\left(\omega\right)\boldsymbol{x}^t\left(\omega\right) + \mathcal{Q}^{t+1}\left(\boldsymbol{x}^t\right) \tag{11.14a}$$

$$\text{s.t. } \boldsymbol{W}^t\boldsymbol{x}^t\left(\omega\right) = \boldsymbol{h}^t\left(\omega\right) - \boldsymbol{T}^{t-1}\left(\omega\right)\boldsymbol{x}^{t-1} \tag{11.14b}$$

$$\boldsymbol{x}^t\left(\omega\right) \geq 0 \tag{11.14c}$$

where $\boldsymbol{x}^t$ indicates the state of the system. Other state information in terms of the realizations of the random parameters up to time t should be included if the distribution of ξ^t is not independent of the past outcomes.

11.2 Chance-Constrained Programming

With random parameters in the optimization problem, we have to determine the decisions prior to the realization of the random parameters. Due to the random effects related to the realizations of the random parameters, we can hardly select the decisions without constraint violation. In the two-stage SP problem, such constraint violation can be handled with the compensations in the second stage, e.g. as done in the maintenance problem of the components in Example 11.2 in Section 11.1. However, for some cases, e.g. safety constraints, compensations do not exist and the constraint violation is almost never avoidable. In such situation, the chance-constrained programming

[13,14] is considered whereby the constraint violation is restricted to a low percentage:

$$\min\{f(\boldsymbol{x})|\mathrm{P}(g(\boldsymbol{x},\xi)\geq 0)\geq p\}, \quad p\in[0,1] \tag{11.15}$$

where, $\mathbf{x}$ is the decision variable vector and ξ is the random parameter vector. The value $p\in[0,1]$ is called probability level. The DM should ensure that the probability of the constraint being satisfied is larger than p. The chance constraint is

$$\mathrm{P}(g(\boldsymbol{x},\xi)\geq 0)\geq p, \quad p\in[0,1] \tag{11.16}$$

which can be rewritten as

$$\alpha(\boldsymbol{x})\geq p, \text{ where } \alpha(\boldsymbol{x}):=\mathrm{P}(g(\boldsymbol{x},\xi)\geq 0) \tag{11.17}$$

The chance-constrained model is often difficult to solve. The main difficulty of chance-constrained programming is that the function $\alpha(\cdot)$ cannot be expressed explicitly in few situations. The theoretical properties and the solution methods are strongly related to the characterizations of the constraint and the random parameters. Therefore, in this chapter, we only consider the chance constraint under three special conditions:

1) Distribution of the random parameters (e.g. continuous, discrete, independent, dependent)
2) Type of constraint system (e.g. linear, separable, coupled)
3) Type of chance constraints (individual, joint)

11.2.1 Model and Properties

i) General chance constraints
The chance constraint in Equation (11.16) can be written more explicitly considering the type of the chance constraints. The first one is to take the probability over the whole constraint system, which is called a joint chance constraint:

$$\mathrm{P}(g_j(\boldsymbol{x},\xi)\geq 0, j=1,\ldots,m)\geq p \tag{11.18}$$

On the other hand, the probability can be considered for each constraint individually:

$$\mathrm{P}(g_j(\boldsymbol{x},\xi)\geq 0)\geq p_j,, j=1,\ldots,m \tag{11.19}$$

This type of individual constraint scheme may yield a large number of inequalities. Comparing to the single constraint in the joint case, this may be mathematically more tractable to solve.

ii) Linear type

When the chance constraints are linear for the random vector, Equation (11.16) can be reformulated as

$$\text{Type I(separated model)} \quad g(\boldsymbol{x},\xi)=h(\boldsymbol{x})-\boldsymbol{A}\xi \tag{11.20}$$

$$\text{Type II(bilinear model)} \quad g(\boldsymbol{x},\xi)=\boldsymbol{A}(\xi)h(\boldsymbol{x})-\boldsymbol{b} \tag{11.21}$$

where $h(\cdot)$ is a function only related to $\boldsymbol{x}$, $\boldsymbol{b}$ is a deterministic vector, $\boldsymbol{A}$ is a deterministic matrix, and $\boldsymbol{A}(\boldsymbol{\xi})$ is a stochastic matrix of $\boldsymbol{\xi}$.

iii) Random right-hand side

The random right-hand side is a special case of the linear separated model in Equation (11.20), with parameter matrix $\boldsymbol{A}$ reduced to the identity matrix. Therefore, the formulation in Equation (11.17) can be given by

$$\alpha(\boldsymbol{x}) = \mathrm{P}(h(\boldsymbol{x}) \geq \boldsymbol{\xi}) = F_{\xi}(h(\boldsymbol{x})) \tag{11.22}$$

where F_{ξ} is the cumulative multivariable distribution function of the random vector ξ. This formulation can be described as the composition formula $\alpha = F_{\xi} \circ h$, and thus, the properties like continuity, convexity and differentiability can be considered.

The model of individual chance constraints in Equation (11.19) with the random right-hand side is given by

$$\alpha_j(\boldsymbol{x}) = \mathrm{P}(h_j(\boldsymbol{x}) \geq \xi_j) = F_{\xi_j}(h_j(\boldsymbol{x})) \geq p_j, j = 1,\ldots,m \tag{11.23}$$

where $\boldsymbol{F}_{\xi_j}$ denotes the one-dimensional cumulative distribution function of the random parameter $\boldsymbol{\xi}_j$. This formula can be inverted by the quantile $\boldsymbol{h}_j(\boldsymbol{x}) \geq \boldsymbol{q}_j(\boldsymbol{p}_j), j = 1,\ldots,\boldsymbol{m}$ where $\boldsymbol{q}_j(\boldsymbol{p}_j)$ is the $\boldsymbol{p}_j$-quantile of $\boldsymbol{F}_{\xi_j}$.

When the components $\{\xi_j\}$ of the random vector ξ are independent, the model of joint chance constraint in Equation (11.18) with the random right-hand side is given by

$$\alpha(\boldsymbol{x}) = \mathrm{P}(h(\boldsymbol{x}) \geq \boldsymbol{\xi}) = F_{\xi_1}(h_1(\boldsymbol{x}))\cdots F_{\xi_m}(h_m(\boldsymbol{x})) \geq p \; j = 1,\ldots,m \tag{11.24}$$

Although this formulation in Equation (11.24) cannot be expressed explicitly like the individual chance constraint model, the one-dimensional cumulative distributions are tractable.

iv) Convexity

The convexity of the feasible set of the chance-constrained programming is essential because this property is a basic issue for any optimization problem:

$$\{\boldsymbol{x} \mid P(g(\boldsymbol{x}, \xi) \geq 0) \geq p\} \tag{11.25}$$

The feasible set of the linear chance constraint with random right-hand side is given by

$$\{\boldsymbol{x} \mid P(h(\boldsymbol{x}) \geq \xi) \geq p\} = \{\boldsymbol{x} \mid F_{\xi}(h(\boldsymbol{x})) \geq p\} \tag{11.26}$$

When the composition function $F_{\xi} \circ h$ is concave, this feasible set is convex. According to the operations to preserve concavity of functions,

1) $F_{\xi} \circ h$ is concave if F_{ξ} is concave and non-decreasing in each argument, and h_j are concave.
2) $F_{\xi} \circ h$ is concave if F_{ξ} is concave and non-increasing in each argument, and h_j are convex.

The cumulative distribution function (cdf) F_{ξ} is non-decreasing. However, it can never be concave due to its bound between 0 and 1. Therefore, we can find a function φ that guarantees the composition $\varphi \circ F_{\xi} \circ h$ is a concave function. Then $\varphi \circ F_{\xi} \circ h$ is concave if $\varphi \circ F_{\xi}$ is concave and non-decreasing in each argument, and h_j are concave. The function

φ can be the function $\log(\cdot)$, since most of the prominent multivariate distribution functions are log-concave. With the convexity of the feasible set, joint chance-constrained programming might be solved with convex optimization methods. For more advanced knowledge about chance-constrained programming, the readers can refer to [13,14].

11.2.2 Example

There are three subsystems i in the system, and each subsystem has one different type of component. The number of the components of each subsystem are a_1, a_2, a_3, respectively. After the system works for a period of time T, the proportion of these three types that has failed is $r=(r_1,r_2,r_3)^T$. To repair all the failed components in the three subsystems, three types of maintainers are available in the market to select, each of which has different maintenance capabilities within the given maintenance time. The maintenance capability of type j maintainer for subsystem i is denoted as b_{ij}. The cost of type j maintainer is c_j per person and in arbitrary units of cost. The decision vector is the number of the maintainers for type j and is indicated as x_j.

Deterministic model:

For the deterministic model, the problem is formulated as follows:

$$\min \sum_{j=1}^{3} c_j x_j$$

$$\boldsymbol{Bx} \geq \boldsymbol{Ar}$$

$$x_j \in Z_+, j=1,2,3$$

where $\boldsymbol{A}$ is a diagonal matrix of a_1, a_2, a_3 and B is the matrix of b_{ij}. We set $c=(4,3,5)$, $r=(0.3,0.5,0.4)$, $a=(100,50,80)$ and

$$B=\begin{pmatrix} 3 & 2 & 0 \\ 2 & 0 & 2 \\ 0 & 2 & 3 \end{pmatrix}$$

The optimal value is 77 with the optimal solution $x=(6,6,7)$.

Chance-constrained model:

To consider the uncertainty of the failure rate $r=(r_1,r_2,r_3)^T$, we look at the individual chance-constrained problem. Assume the failure rate for type i denoted by ξ_i follows the normal distribution, i.e. $\xi_i \sim N(r_i, \sigma_i^2)$. The chance-constrained model is

$$\min \sum_{j=1}^{3} c_j x_j$$

$$\mathrm{P}\left(\sum_{j=1}^{3} b_{ij} x_j \geq a_i \xi_i\right) \geq p, i=1,2,3$$

$$x_j \in Z_+, j=1,2,3$$

The chance constraints can be rewritten as $\sum_{j=1}^{3} b_{ij}x_j \geq a_i r_i + a_i \sigma_i q_p, i = 1,2,3$ where q_p is the p-quantile of the standard normal distribution. Set the probability level p as 0.9 and σ_i as 0.1, 0.2, 0.3 respectively. We can solve this problem as an integer program. The solution is $x = (6,13,13)$ and the optimal value is 128.

11.3 Robust Optimization (RO)

The stochastic optimization and chance-constrained problem illustrated in the previous sections mainly deal with the uncertain parameters when their probability distributions are known. When the probability distribution of an uncertain parameter is unknown and the uncertain parameter values are known to reside in the uncertainty set, robust optimization (RO) [15–17] can be considered. RO guarantees the feasibility of all constraints under any realization of the parameters within the uncertainty set. The original RO dates back to the 1940s, using worst-case analysis and Wald's maximin model [18] as tools to treat severe uncertainty.

Suppose uncertainty exists in the objective function. The uncertainty parameters $\boldsymbol{u} \in R^k$ are assumed to take arbitrary values in the uncertainty set $\mathcal{U} \subseteq R^k$, and the problem can be formulated as follows:

$$\begin{aligned} &\min f_0(\boldsymbol{x},\boldsymbol{u}) \\ &s.t.\ f_i(\boldsymbol{x}) \leq 0, i = 1,\ldots,m \end{aligned} \tag{11.27}$$

where $\boldsymbol{x} \in R^n$ is a vector of decision variables, and $f_0, f_i : R^n \to R$ are functions. The min-max and min-max regret criteria are often used to hedge against parameters variations. The min-max criterion aims to obtain a solution that achieves the best possible performance in the worst case. The min-max regret criterion, less conservative, aims at obtaining a solution minimizing the maximum deviation between the value of the solution and the optimal value of the corresponding uncertainty value over all possible uncertainty values.

The min-max version considers to find a solution under the worst-case value across all $\boldsymbol{u} \in \mathcal{U}$, which is given by

$$\min\left\{\max_{u \in \mathcal{U}} f_0(\boldsymbol{x},\boldsymbol{u}) : f_i(\boldsymbol{x}) \leq 0, i = 1,\ldots,m\right\}$$

Given the feasible solution x, its regret under the uncertainty value $\boldsymbol{u} \in \mathcal{U}$ is defined as

$$Reg(\boldsymbol{x},\boldsymbol{u}) = f_0(\boldsymbol{x},\boldsymbol{u}) - f_0(\boldsymbol{x}_u^*,\boldsymbol{u})$$

where $\boldsymbol{x}_u^*$ is an optimal solution under the uncertainty parameter $\boldsymbol{u}$ and $f_0(\boldsymbol{x}_u^*,\boldsymbol{u})$ is the corresponding optimal value. The min-max regret version considers finding a solution minimizing its maximum regret, which is given by

$$\begin{aligned} &\min\left\{Reg_{max}(\boldsymbol{x}) : f_i(\boldsymbol{x}) \leq 0, i = 1,\ldots,m\right\} \\ &\quad = \min\left\{\max_{u \in \mathcal{U}}\left(f_0(\boldsymbol{x},\boldsymbol{u}) - f_0(\boldsymbol{x}_u^*,u)\right) : f_i(\boldsymbol{x}) \leq 0, i = 1,\ldots,m\right\} \end{aligned}$$

In this section, we introduce the RO under the uncertain linear optimization (LO) problem to show its properties.

11.3.1 Uncertain Linear Optimization (LO) and its Robust Counterparts

Definition 11.1 An *uncertain LO problem* is a collection

$$LO_{\mathcal{U}} = \left\{ \min_{x} \left\{ \mathbf{c}^T \boldsymbol{x} + d : \boldsymbol{Ax} \leq \mathbf{b} \right\} \right\}_{(c,d,\boldsymbol{A},\mathbf{b}) \in \mathcal{U}} \tag{11.28}$$

of general LO problems $\min_x \left\{ c^T x + d : \boldsymbol{Ax} \leq \mathbf{b} \right\}$, which includes m constraints and n variables with the data $(\boldsymbol{c}, d, \boldsymbol{A}, \boldsymbol{b})$ varying in a given *uncertainty set* $\mathcal{U} \subset R^{(m+1)\times(n+1)}$. We often assume that the uncertainty set is parameterized in an affine fashion with *perturbation vector* ζ varying in a given *perturbation set* $\mathcal{Z}$:

$$\mathcal{U} = \left\{ \begin{pmatrix} \mathbf{c}^T & d \\ \boldsymbol{A} & \mathbf{b} \end{pmatrix} = \begin{pmatrix} \mathbf{c}_0^T & d_0 \\ \boldsymbol{A}_0 & \mathbf{b}_0 \end{pmatrix} + \sum_{l=1}^{L} \zeta_l \begin{pmatrix} \mathbf{c}_l^T & d_l \\ A_l & \mathbf{b}_l \end{pmatrix} : \zeta \in \mathcal{Z} \subset R^L \right\}$$

Definition 11.2 A vector $\boldsymbol{x}$ is a *robust feasible solution* to $LO_{\mathcal{U}}$ if it satisfies the constraints for any realization of uncertain data from the uncertainty set, i.e.

$$\boldsymbol{Ax} \leq b, \forall (\mathbf{c}, d, \boldsymbol{A}, \mathbf{b}) \in \mathcal{U}$$

Definition 11.3 Given a robust feasible solution $\boldsymbol{x}$, the *robust value* $\hat{c}(\boldsymbol{x})$ of the objective in $LO_{\mathcal{U}}$ is the largest value of objective $\mathbf{c}^T \boldsymbol{x} + d$ over all realizations of the uncertain data, i.e.

$$\hat{c}(\boldsymbol{x}) = \sup_{(c,d,\boldsymbol{A},\mathbf{b}) \in \mathcal{U}} \left(\mathbf{c}^T \boldsymbol{x} + d \right)$$

Definition 11.4 The *robust counterpart* (RC) of the uncertain problem $LO_{\mathcal{U}}$ is the optimization problem which minimizes the robust value of the objective over all robust feasible solutions, that is,

$$\min_{x} \left\{ \hat{c}(\mathbf{x}) = \sup_{(c,d,\boldsymbol{A},\mathbf{b}) \in \mathcal{U}} \{ \left(c^T \mathbf{x} + d \right) : \boldsymbol{Ax} \leq \mathbf{b} \right\}$$

The optimal solution of RC is called a robust optimal solution to $LO_{\mathcal{U}}$, and the corresponding objective value is called the robust optimal value of $LO_{\mathcal{U}}$. Here are some properties of the uncertain LO problem. For details of the general proof of these properties, see [15].

Remark 11.1 An uncertain LO problem can always be translated into an uncertain LO problem with certain objective. W.l.o.g.,[1] we can restrict the uncertain LO problem with certain objectives.

---

1 Abbr. for "without loss of generality".

Remark 11.2 If the right-hand side of the constraint is uncertain, we can translate these uncertain data by adding a new variable $x_{n+1} = -1$, whose coefficient is this uncertain data. W.l.o.g., we can restrict the uncertain LO problem with certain right-hand side constraints.

Remark 11.3 The uncertainty set $\mathcal{U}$ can be replaced by its convex hull $conv(\mathcal{U})$.

Remark 11.4 The uncertainty in the data can be modelled constraint-wise. Assume that $LO_{\mathcal{U}}$ is with certain objective. Then the RC of $LO_{\mathcal{U}}$ is

$$\min_x \left\{ \boldsymbol{c}^T \boldsymbol{x} + d : \boldsymbol{A}\boldsymbol{x} \leq \boldsymbol{b}, \forall (\boldsymbol{A}, \boldsymbol{b}) \in \mathcal{U} \right\}$$

If we consider each constraint $(\boldsymbol{Ax})_i \leq b_i$, then

$$a_i^T \boldsymbol{x} \leq b_i, \forall (\boldsymbol{a}_i, b_i) \in \mathcal{U}_i$$

where $\boldsymbol{a}_i^T$ is the i-th row of $\boldsymbol{A}$ and $\mathcal{U}_i$ is the projection of $\mathcal{U}$ on the i-th constraint. The RC of $LO_{\mathcal{U}}$ with a certain objective remains intact when the uncertainty set $\mathcal{U}$ is extended to the direct product $\mathcal{U} = \mathcal{U}_1 \times \ldots \times \mathcal{U}_m$.

11.3.2 Tractability of Robust Counterparts

According to the remarks mentioned above, w.l.o.g., we consider the uncertain LO problem with a certain objective, certain right-hand side, and a single constraint because of the constraint-wise property

$$\left\{ \boldsymbol{a}^T \boldsymbol{x} \leq b \right\}_{a \in \mathcal{U}} \tag{11.29}$$

The data varying in the uncertainty set are

$$\mathcal{U} = \left\{ \boldsymbol{a} = \boldsymbol{a}^0 + \sum_{l=1}^{L} \zeta_l \boldsymbol{a}^l = \boldsymbol{a}^0 + \boldsymbol{D}\zeta : \zeta \in \mathcal{Z} \right\}$$

where $\boldsymbol{D} \in R^{n \times L}$. Assume that the perturbation set $\mathcal{Z}$ is convex.

Equation (11.29) contains infinite constraints due to the perturbation vector on set $\mathcal{Z}$, and it seems intractable in this formulation. The goal is to build a representation to reformulate this semi-infinite linear constraint as a finite system of explicit convex constraints and to convert the RC of $LO_{\mathcal{U}}$ into an explicit and tractable convex program.

A single constraint in Equation (11.29) equals to

$$\left(\boldsymbol{a}^0 + \boldsymbol{D}\zeta \right)^T \boldsymbol{x} \leq b, \forall \zeta \in \mathcal{Z} \tag{11.30}$$

We consider that the perturbation set $\mathcal{Z}$ is polyhedral:

$$\mathcal{Z} = \left\{ \zeta : \boldsymbol{P}\zeta + \boldsymbol{q} \geq 0 \right\}$$

where $\boldsymbol{P} \in R^{h \times L}$, $\xi \in R^L$, and $q \in R^h$. Therefore, Equation (11.30) can be converted into

$$(11.30) \Leftrightarrow \left(\boldsymbol{a}^0 \right)^T \boldsymbol{x} + \max_{\zeta : P\zeta + q \geq 0} \left(\boldsymbol{D}^T \boldsymbol{x} \right)^T \zeta \leq b$$

$$\Leftrightarrow \left(\boldsymbol{a}^0 \right)^T \boldsymbol{x} + \min_{w} \left\{ \boldsymbol{q}^T \boldsymbol{w} : \boldsymbol{P}^T \boldsymbol{w} = -\boldsymbol{D}^T \boldsymbol{x}, \boldsymbol{w} \geq 0 \right\} \leq b$$

$$\Leftrightarrow \exists \boldsymbol{w} : \left(\boldsymbol{a}^0 \right)^T \boldsymbol{x} + \boldsymbol{q}^T \boldsymbol{w} \leq b, \boldsymbol{P}^T \boldsymbol{w} = -\boldsymbol{D}^T \boldsymbol{x}, \boldsymbol{w} \geq 0$$

The second equality uses the strong duality of LO. All constraints and the objective are linear, and the RC with this representation is tractable. Table 11.2 shows the tractable RC representations of an uncertain LO problem for different perturbation sets Z. [19].

11.3.3 Robust Optimization (RO) with Cardinality Constrained Uncertainty Set

The robust approaches, i.e. box and ellipsoidal uncertainty sets, are too conservative [20]. Reference [21] proposed the cardinality constrained uncertainty to control the robustness to withstand parameter uncertainty. We consider the constraint in Equation (11.29) with the uncertainty set as

$$\mathcal{U} = \left\{ \boldsymbol{a} : a_i \in \left[a_i^0 - \hat{a}_i, a_i^0 + \hat{a}_i \right], i \in I \right\}$$

where I is the index set of all variables x_i and a_i^0 is the nominal value of uncertain data a_i. The range of variation on uncertain data a_i is $\hat{a}_i$ and a_i takes values according to a symmetric distribution in interval $\left[a_i^0 - \hat{a}_i, a_i^0 + \hat{a}_i \right]$. The parameter $\Gamma \in \left[0, |I| \right]$ is introduced to adjust the conservative level of the robust solution. It is unlikely that all uncertain parameters will change, i.e., up to $\lfloor \Gamma \rfloor$ of all uncertain parameters are allowed to change by $\hat{a}_i$ and one parameter is allowed to change by $\left(\Gamma - \lfloor \Gamma \rfloor \right) \hat{a}$. Then the constraint in Equation (11.29) is formulated by

$$\sum_{i \in I} a_i^0 x_i + \sum_{i \in S} \hat{a}_i x_i + \left(\Gamma - \Gamma \right) \hat{a}_s x_s \le b, \ \forall \left\{ S \cup \{s\} : S \subseteq I, |S| = \lfloor \Gamma \rfloor, s \in I \setminus S \right\} \quad (11.31)$$

which is equivalent to

$$\sum_{i \in I} a_i^0 x_i + \max_{\{S \cup \{s\} : S \subseteq I, |S| = \lfloor \Gamma \rfloor, s \in I \setminus S\}} \left\{ \sum_{i \in S} \hat{a}_i x_i + \left(\Gamma - \lfloor \Gamma \rfloor \right) \hat{a}_s x_s \right\} \le b$$

Table 11.2 Tractable RC representations given different Z.

Perturbation set type	Z	RC	Tractability
Box	$\lVert \zeta \rVert_\infty \le 1$	$\left(\boldsymbol{a}^0 \right)^T \boldsymbol{x} + \lVert \boldsymbol{D}^T \boldsymbol{x} \rVert_1 \le \boldsymbol{b}$	LP
Ellipsoidal	$\lVert \zeta_2 \rVert \le 1$	$\left(\boldsymbol{a}^0 \right)^T \boldsymbol{x} + \lVert \boldsymbol{D}^T \boldsymbol{x}_2 \rVert \le \boldsymbol{b}$	CQP
Polyhedral	$\boldsymbol{P}\zeta + \boldsymbol{q} \ge 0$	$\begin{cases} \left(\boldsymbol{a}^0 \right)^T \boldsymbol{x} + \boldsymbol{q}^T \boldsymbol{w} \le \boldsymbol{b} \\ \boldsymbol{P}^T \boldsymbol{w} = -\boldsymbol{D}^T \boldsymbol{x} \\ \boldsymbol{w} \ge 0 \end{cases}$	LP
Cone	$\boldsymbol{P}\zeta + \boldsymbol{q} \in \boldsymbol{K}$	$\begin{cases} \left(\boldsymbol{a}^0 \right)^T \boldsymbol{x} + \boldsymbol{q}^T \boldsymbol{w} \le \boldsymbol{b} \\ \boldsymbol{P}^T \boldsymbol{w} = -\boldsymbol{D}^T \boldsymbol{x} \\ \boldsymbol{w} \in \boldsymbol{K}^* \end{cases}$	Conic opt.

The constraint is intractable because the combinations of set $\{S \cup \{s\}\}$ are exponential under the operation max. We give the tractable representations as follows:

Given $\boldsymbol{x}$, we define

$$\phi(x) = \max_{\{S\cup\{s\}: S\subseteq I, |S|=\lfloor\Gamma\rfloor, s\in I\setminus S\}} \left\{ \sum_{i\in S} \hat{a}_i x_i + \left(\Gamma - \lfloor\Gamma\rfloor\right)\hat{a}_s x_s \right\} \uplus$$

This equals

$$\phi(x) = \max \sum_{i\in I} \hat{a}_i x_i \psi_i$$

$$\text{s.t. } \sum_{i\in I} \psi_i \le \Gamma$$

$$0 \le \psi_i \le 1, \forall i \in I$$

$\phi(x)$ is equivalent to the following problem using the strong duality of LP:

$$\phi(x) = \min \quad \xi\Gamma + \sum_{i\in I} \rho_i$$

$$\text{s.t. } \xi + \rho_i \ge \hat{a}_i x_i, \forall i \in I$$

$$\rho_i \ge 0, \forall i \in I$$

$$\xi \ge 0$$

Then the tractable reformulations of Equation (11.31) are given by

$$\xi + \sum_{i\in I} \rho_i \le b \tag{11.32}$$

$$\xi + \rho_i \ge \hat{a}_i x_i, \forall i \in I \tag{11.33}$$

$$\rho_i \ge 0, \forall i \in I \tag{11.34}$$

$$\xi \ge 0 \tag{11.35}$$

When $\Gamma^c = 0$, the uncertainty of parameter $\boldsymbol{a}$ is not considered in the constraint in Equation (11.31). When $\Gamma^c = |I|$, the most conservative formulation of the uncertain data is considered.

11.3.4 Example

In this section, we give an example to illustrate the RO applied to reliability optimization problems. We consider a problem similar to Example 1 in Section 11.1. The difference is that the repairmen cost is certain and equal to 3.5 per unit, and instead of knowing the probability distribution of the uncertain survival rate $\boldsymbol{a}$, we know that $\boldsymbol{a}$ resides in the uncertainty set $\mathcal{U}=\{\boldsymbol{a}^0+\boldsymbol{D}\zeta:\zeta\in Z\}$ and $Z=\{\zeta\in R^3:\zeta_\infty\leq 1\}$. $a^0=(0.7,0.6,0.8)$ and $\boldsymbol{D}=0.1\boldsymbol{E}$ where $\boldsymbol{E}$ is the 3×3 unit matrix. Therefore, the design to minimize the purchase cost and maintenance cost is

$$\min 0.4x_1+0.8x_2+0.7x_3+3.5y$$

$$\boldsymbol{a}^T\boldsymbol{x}+y\geq 0.9,\forall a\in\mathcal{U}$$

$$\boldsymbol{e}^T\boldsymbol{x}=1$$

$$x,y\geq 0$$

where $\boldsymbol{e}=(1,1,1)^T$. The first constraint ensures that 90% of components are working after the maintenance. The second constraint means that the total percentage of all components is 1. In this case, RC can be formulated as

$$\begin{aligned}
\text{RC}&\Leftrightarrow \boldsymbol{a}^T\boldsymbol{x}+y\geq 0.9,\forall \boldsymbol{a}\in\mathcal{U}\\
&\Leftrightarrow \left(\boldsymbol{a}^0\right)^T\boldsymbol{x}+\min_{\zeta:\|\zeta\|_\infty\leq 1}\left(\boldsymbol{D}^T\boldsymbol{x}\right)^T\zeta\geq 0.9\\
&\Leftrightarrow \left(\boldsymbol{a}^0\right)^T\boldsymbol{x}-\left\|\boldsymbol{D}^T\boldsymbol{x}\right\|_1\geq 0.9\\
&\Leftrightarrow \begin{cases}-u_l\leq\left(\boldsymbol{d}^l\right)^T\boldsymbol{x}\leq u_l, & l=1,\ldots,L\\ \left(\boldsymbol{a}^0\right)^T\boldsymbol{x}-\sum_{l=1}^{L}u_l\geq 0.9\end{cases}
\end{aligned}$$

where $\boldsymbol{d}^l$ is the l-th column of the matrix $\boldsymbol{D}$.

Therefore, RC can be represented by a tractable representation as follows:

$$\min 0.4x_1+0.8x_2+0.7x_3+3.5y$$

$$0.1x_1\leq u_1$$

$$0.1x_2\leq u_2$$

$$0.1x_3\leq u_3$$

$$0.7x_1+0.6x_2+0.8x_3+y-u_1-u_2-u_3\geq 0.9$$

$$\boldsymbol{e}^T\boldsymbol{x}=1$$

$$\boldsymbol{x},y,\boldsymbol{u}\geq 0$$

The robust optimal solution is $\boldsymbol{x}^*=(0,0,1)$ and $y^*=0.2$. The robust optimal value of objective is 1.4.

11.4 Exercises

1. Solve the following two-stage SP problem by L-shaped method:

$$z = \min 100x_1 + 150x_2 + E_{\xi}\left(q_1y_1 + q_2y_2\right)$$

$$x_1 + x_2 \leq 120$$

$$7y_1 + 10y_2 \leq 60x_1$$

$$6y_1 + 5y_2 \leq 80x_2$$

$$y_1 \leq \boldsymbol{d}_1,\ y_2 \leq \boldsymbol{d}_2$$

$$x_1 \geq 40,\ x_2 \geq 20,\ y_1, y_2 \geq 0$$

where $\xi^T = \left(d_1, d_2, q_1, q_2\right)$ takes values (450, 100, -24, -28) with probability 0.7 and (300, 400, -25, -30) with probability 0.3.

2. Consider the example in Section 11.2.2 but now with ξ following the uniform distribution, i.e. $\xi_i \sim U\left[r_i - \hat{r}_i, r_i - \hat{r}_i\right]$ and $\hat{r}_i = r_i / 4$. Show that the chance-constrained model follows the same path as before.
3. Consider the example in Section 11.3.3 but now with the uncertainty set $\mathcal{U} = \left\{\boldsymbol{a}^0 + \boldsymbol{D}\zeta : \zeta \in Z\right\}$ and $Z = \left\{\zeta \in R^3 : \zeta_2 \leq 2\right\}$. Give the RC representation of this problem and solve it.
4. Consider the redundancy allocation problem (RAP) for binary-state series-parallel system:

$$\max \prod_{j \in J}\left(1 - \left(1 - r_j\right)^{x_j}\right)$$

$$\boldsymbol{A}x \leq \boldsymbol{b}$$

$$l \leq x \leq u$$

$$x \in Z_+^n$$

where $J = \left\{1, 2, \ldots, n\right\}$ and $A \in R^{m \times n}$. Suppose the component reliability $\tilde{r}_j$ in subsystem $j \in J$ is uncertain, and it takes a random value in $\left[\hat{r}_j - \delta_j, \hat{r}_j\right]$: that is, $\tilde{r}_j = \hat{r}_j - \delta_j \xi_j$ where the perturbations ξ_j are n independent random variables with $0 \leq \xi_j \leq 1$. Show the tractable RC representation for this reliability optimization.

5. Budget uncertainty set [21] is a less conservative approach than the box uncertainty set for the robust problem. For Exercise 4, consider the budget uncertainty set instead of the box uncertainty set, i.e.

$$\sum_{j \in J} \xi_j \leq \Gamma$$

where $\Gamma \in \left[0, n\right]$ and is not necessarily an integer. The role of Γ is to adjust the robustness of the model against the level of conservatism of the solution. Give the tractable RC representation for this reliability optimization.

References

1 Birge, J.R. and Louveaux, F. (2011). *Introduction to Stochastic Programming*. New York City: Springer Science & Business Media.
2 Dantzig, G.B. (2010). Linear programming under uncertainty. In: *Stochastic Programming* 50 (12 supplement, 1764–1769. New York City: Springer.
3 Beale, E.M. (1955). On minimizing a convex function subject to linear inequalities. *Journal of the Royal Statistical Society: Series B (Methodological)* 17 (2): 173–184.
4 Dantzig, G.B. and Glynn, P.W. (1990). Parallel processors for planning under uncertainty. *Annals of Operations Research* 22 (1): 1–21.
5 Higle, J.L. and Sen, S. (1991). Stochastic decomposition: An algorithm for two-stage linear programs with recourse. *Mathematics of Operations Research* 16 (3): 650–669.
6 Higle, J.L. and Sen, S. (1996). *Stochastic Decomposition: A Statistical Method for Large Scale Stochastic Linear Programming*. Secaucus, NJ: Springer Science & Business Media.
7 Ermoliev, Y.M. and Wets, R.-B. (1988). *Numerical Techniques for Stochastic Optimization*. Heidelberg, Germany: Springer-Verlag.
8 Van Slyke, R.M. and Wets, R. (1969). L-shaped linear programs with applications to optimal control and stochastic programming. *SIAM Journal on Applied Mathematics* 17 (4): 638–663.
9 Rahmaniani, R., Crainic, T.G., Gendreau, M., and Rei, W. (2017). The Benders decomposition algorithm: A literature review. *European Journal of Operational Research* 259 (3): 801–817.
10 Vanderbeck, F. and Savelsbergh, M.W. (2006). A generic view of Dantzig-Wolfe decomposition in mixed integer programming. *Operations Research Letters* 34 (3): 296–306.
11 Zahiri, B., Torabi, S.A., Mohammadi, M., and Aghabegloo, M. (2018). A multi-stage stochastic programming approach for blood supply chain planning. *Computers Industrial Engineering* 122: 1–14.
12 Yahyatabar, A. and Najafi, A.A. (2018). Condition based maintenance policy for series-parallel systems through Proportional Hazards Model: A multi-stage stochastic programming approach. *Computers Industrial Engineering* 126: 30–46.
13 Dentcheva, D. (2006). Optimization models with probabilistic constraints. In: *Probabilistic and Randomized Methods for Design under Uncertainty* (ed. G. Calafiore and F. Dabbene), 49–97. Springer.
14 Prékopa, A. (2003). Probabilistic programming. *Handbooks in Operations Research and Management Science* 10: 267–351.
15 Ben-Tal, A., El Ghaoui, L., and Nemirovski, A. (2009). *Robust Optimization*. Princeton, NJ: Princeton University Press.
16 Ben-Tal, A. and Nemirovski, A. (2008). Selected topics in robust convex optimization. *Mathematical Programming* 112 (1): 125–158.
17 Bertsimas, D., Brown, D.B., and Caramanis, C. (2011). Theory and applications of robust optimization. *SIAM Review* 53 (3): 464–501.

18 Wald, A. (1945). Statistical decision functions which minimize the maximum risk. In: *Annals of Mathematics*, 265–280.

19 Gorissen, B.L., Yanıkoğlu, İ., and Den Hertog, D. (2015). A practical guide to robust optimization. *Omega* 53: 124–137.

20 Bertsimas, D. and Sim, M. (2004). The price of robustness. *Operations Research* 52 (1): 35–53.

21 Bertsimas, D. and Sim, M. (2003). Robust discrete optimization and network flows. *Mathematical Programming* 98 (1): 49–71.

12

Applications

This chapter contains two application cases that make use of the optimization methods introduced in the previous chapters of Part III. The first case study considers optimizing the design of a distributed power generation system under various uncertainties. Multi-objective optimization (MOO) and Monte Carlo simulation (MCS) are implemented to solve this problem. The second case study is about redundancy allocation for binary-state series-parallel systems (BSSPSs) under epistemic uncertainty.

12.1 Multi-objective Optimization (MOO) Framework for the Integration of Distributed Renewable Generation and Storage

We present a MOO framework for integrating renewable generators and storage devices into an electrical distribution network. The framework searches for the optimal size and location of the distributed renewable generation units. Uncertainties in renewable resources availability, components failure and repair events, loads and grid power supply are incorporated. A Monte Carlo simulation – optimal power flow (MCS-OPF) computational model is used to generate scenarios of the uncertain variables and evaluate the network electric performance. For monitoring and controlling the risk associated to the performance of the distributed generation (DG) system, we consider the conditional value-at-risk (CVaR) measure within the framework. The MOO problem is formulated with respect to the minimization of the expectations of the global cost (C_g) and Energy Not Supplied (ENS), combined with their respective CVaR. The fast non-dominated sorting genetic algorithm (NSGA-II) [1] is used for the MOO framework. The framework is applied to a distribution network derived from the IEEE 13 nodes test feeder [2].

Reliability Analysis, Safety Assessment and Optimization: Methods and Applications in Energy Systems and Other Applications, First Edition. Enrico Zio and Yan-Fu Li.

12.1.1 Description of Distributed Generation (DG) System

The DG system model, presented in Section 7.1.1, has neglected many of the topological and electrical characteristics of the DG system because the adequacy assessment generally does not require such information. However, in this chapter we intend to introduce a more detailed DG system model for a better approximation to the real-world DG system and will obtain practical optimization results for the allocation of the DG generators.

Four main classes of components are considered: nodes, feeders, renewable DG units and main supply power spots (MSs). The nodes can be understood as fixed spatial locations at which generation units and loads can be allocated. Feeders connect different nodes and through them the power is distributed. Renewable DG units and MSs are power sources; for electric vehicles (EVs) and storage devices, they can also act as loads when they are in charging state. The locations of the MSs are fixed. The MOO aims at optimally allocating renewable DG units at the different nodes. Figure 12.1 shows an example of configuration of a DG system adapted from the IEEE 13 nodes test feeder [3], where the regulator, capacitor, switch, and the feeders with length equal to zero are neglected.

The renewable DG technologies include solar photovoltaic (PV), wind turbines (W), electric vehicles (EV) and storage devices (ST), i.e. batteries. The power output of each of these technologies is inherently uncertain. PV and W generations are subject to variability through their dependence on environmental conditions, i.e., solar irradiance and wind speed. Dis/connection and dis/charging patterns in EV and ST, respectively, further influence the uncertainty in the power outputs from the DG units. Also generation and distribution interruptions caused by failures are regarded as significant. The details about different types of DG unit models can be found in publication [4].

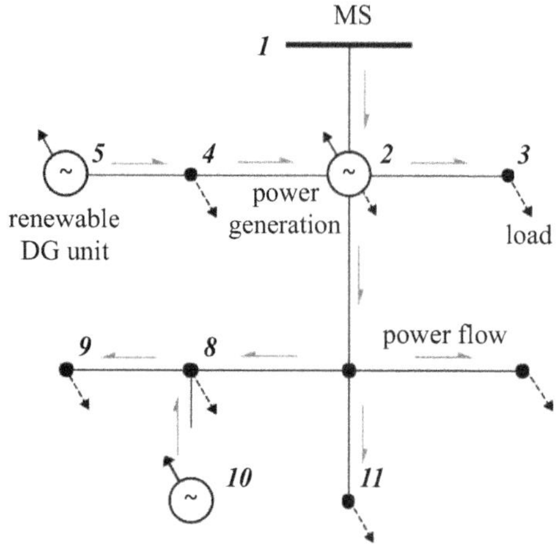

Figure 12.1 Example of distribution generation (DG) system configuration [4].

We will introduce the basic notations of this application case, as follows:

$\mathcal{N}$	set of all nodes
$\mathcal{M}$	set of all types of main supply power sources
$\mathcal{D}$	set of all DG technologies
$\mathcal{P}_v$	set of all PV technologies
$\mathcal{W}$	set of all wind technologies
$\mathcal{E}_v$	set of all EV technologies
$\mathcal{S}_T$	set of all ST
$\mathcal{F}$	set of all feeders

The configurations of power sources allocated in the network, indicating the size of power capacity and the location, is given in matrix form:

$$\Xi = \left[\begin{array}{ccccc|ccccc} \xi_{1,1} & \cdots & \xi_{1,j} & \cdots & \xi_{1,\mathcal{M}} & \xi_{1,\mathcal{M}+1} & \cdots & \xi_{1,\mathcal{M}+j} & \cdots & \xi_{1,\mathcal{M}+\mathcal{D}} \\ \vdots & \ddots & \vdots & & \vdots & \vdots & \ddots & \vdots & & \vdots \\ \xi_{i,1} & \cdots & \xi_{i,j} & \cdots & \xi_{i,\mathcal{M}} & \xi_{i,\mathcal{M}+1} & \cdots & \xi_{i,\mathcal{M}+j} & \cdots & \xi_{i,\mathcal{M}+\mathcal{D}} \\ \vdots & & \vdots & \ddots & \vdots & \vdots & & \vdots & \ddots & \vdots \\ \xi_{\mathcal{N},1} & \cdots & \xi_{\mathcal{N},j} & \cdots & \xi_{\mathcal{N},\mathcal{M}} & \xi_{\mathcal{N},\mathcal{M}+1} & \cdots & \xi_{\mathcal{N},\mathcal{M}+j} & \cdots & \xi_{\mathcal{N},\mathcal{M}+\mathcal{D}} \end{array}\right] = \left[\Xi^{\mathcal{M}} \quad \Xi|^{\mathcal{D}}\right] \tag{12.1}$$

where

Ξ	*configuration matrix* of type, size and location of the power sources allocated in the distribution network
$\Xi^{\mathcal{M}}$	allocated main supply part of the configuration matrix
$\Xi^{\mathcal{D}}$	allocated DG units part of the configuration matrix
n	number of nodes in the network, $\lvert\mathcal{N}\rvert$
m	number of main supply type (transformers), $\lvert\mathcal{M}\rvert$
d	number of DG technologies, $\lvert\mathcal{D}\rvert$

$$\xi_{ij} = \begin{cases} \varsigma \text{ number of units of the MS type or DG technology } j \text{ allocated at node } i \\ 0 \quad \text{otherwise} \end{cases} \quad \forall i \in \mathcal{N}, j \in \mathcal{M} \bigcup \mathcal{D}, \varsigma \in Z^{+} \tag{12.2}$$

Feeders deployment is described by the set of the node pairs connected:

$$\mathcal{F} = \{(1,2),\ldots,(i,i')\} \ \forall (i,i') \in \mathcal{N} \times \mathcal{N}, (i,i') \text{ is a feeder}, \tag{12.3}$$

Any configuration $\{\Xi, \mathcal{F}\}$ of power sources Ξ and feeders $\mathcal{F}$ of the distribution network is affected by uncertainty, so the operation and performance of the distribution network is strongly dependent on the network configuration and scenarios.

Non-sequential MCS is adopted to sample the output of each component without time dependence, with the aim of reducing the computation times. For a given structure and configuration of the distribution network $\{\Xi, \mathcal{F}\}$, the set $\vec{\vartheta}$ of sampled output variables constitutes an operational scenario in correspondence of which the distribution network operation is modeled by optimal power flow (OPF) and its performance evaluated. The two inputs to the OPF model are the network configuration $\{\Xi, \mathcal{F}\}$ and the operational conditions scenario $\vec{\vartheta}$:

$$\vec{\vartheta} = \left[t_d, P_{i,j}^{ms}, L_i, s_i, ws_i, t_{Rop_{i,j}}, Q_{i,j}^{st}, mc_{i,j}, mc_{i,i'}\right] \forall i,i' \in \mathcal{N}, j \in \mathcal{M} \cup \mathcal{D}, (i,i') \in \mathcal{F}, \tag{12.4}$$

where,

t_d	hour of the day $[h]$, randomly sampled from a uniform distribution $U(1,24)$
$P_{i,j}^{ms}$	main supply power of the power source j at node i $[kW]$
L_i	power demand at node i $[kW]$
s_i	solar irradiance at node i $[kW/m^2]$
ws_i	wind speed at node i $[m/s]$
$t_{Rop_{i,j}}$	residence time interval for operating state op of the power source j at node i $[h]$
$Q_{i,j}^{st}$	level of charge in the battery in the power source j at node i $[KJ]$
$mc_{i,j}$	binary mechanical state variable of the power source j at node i
$mc_{i,i'}$	binary mechanical state variable of the feeder (i,i')

12.1.2 Optimal Power Flow (OPF)

Power system analysis is performed by direct current (DC) OPF, which takes into account the active power flows, neglecting power losses, and assumes a constant value of the voltage throughout the network. This allows to transform the classical nonlinear power flow formulation into a linear one, gaining simplicity and computational tractability. For a given configuration $\{\Xi, \mathcal{F}\}$ and operational scenario $\vec{\vartheta}$ the formulation of the OPF problem is:

$$\min C_{O\&M}^{net\vartheta}\left(P_{Gu}^{\vec{\vartheta}}\right) = \sum_{i \in N} \sum_{j \in \mathcal{M} \cup \mathcal{D}} C_{O\&M_j^v} \times P_{Gu_{i,j}}^{\vec{\vartheta}} \times t^h \tag{12.5a}$$

$$\text{s.t.} \left(\sum_{j \in \mathcal{M} \cup \mathcal{D}} P_{Gu_{ij}}^{\vec{\vartheta}} + LS_i^{\vec{\vartheta}} + \sum_{i' \in N} mc_{i,i'}^{\vec{\vartheta}} B_{i,i'}\left(\delta_i^{\vec{\vartheta}} - \delta_{i'}^{\vec{\vartheta}}\right)\right) - L_{i'}^{\vec{\vartheta}} = 0 \ \forall i,i' \in \mathcal{N}, (i,i') \in \mathcal{F} \tag{12.5b}$$

$$P_{Gu_{i,j}}^{\vec{\vartheta}} \leq P_{Ga_{i,j}}^{\vec{\vartheta}} \ \forall i \in \mathcal{N}, j \in \mathcal{M} \cup \mathcal{D} \tag{12.5c}$$

$$0 \leq P_{Gu_{i,j}}^{\vec{\vartheta}} \ \forall i \in \mathcal{N}, j \in \mathcal{M} \cup \mathcal{D} \tag{12.5d}$$

$$mc_{i,i'}^{\vec{\vartheta}} B_{i,i'}\left(\delta_i^{\vec{\vartheta}} - \delta_{i'}^{\vec{\vartheta}}\right) \leq V \times Amp_{i,i'} \ \forall i,i' \in \mathcal{N}, (i,i') \in \mathcal{F} \tag{12.5e}$$

$$-mc_{i,i'}^{\vec{\vartheta}} B_{i,i'}\left(\delta_i^{\vec{\vartheta}} - \delta_{i'}^{\vec{\vartheta}}\right) \leq V \times Amp_{i,i'} \ \forall i,i' \in \mathcal{N}, (i,i') \in \mathcal{F} \tag{12.5f}$$

where,

t^h	duration of the scenario $[h]$
$C_{O\&M}^{net\vartheta}$	operating and maintenance costs of the total power supply and generation $[\$]$
$C_{O\&M_j^v}$	operating and maintenance variable costs of the power source $j[\$/kWh]$
$mc_{i,i'}^{\vartheta}$	mechanical state of the feeder (i,i')
$B_{i,i'}$	susceptance of the feeder (i,i') $[1/\Omega]$
$mc_{i,j}^{\vartheta}$	mechanical state of the power source j at node i
$P_{Ga_{i,J}}^{\vartheta}$	available power in the source j at node i $[kW]$
$P_{Gu_{i,J}}^{\vartheta}$	power produced by source j at node i $[kW]$
LS_i^{ϑ}	load shedding at node i $[kW]$
V	nominal voltage of the network $[kV]$
$Amp_{i,i'}$	ampacity of the feeder (i,i') $[A]$

The load shedding in the node i, LS_i, is defined as the amount of load(s) disconnected in node i to alleviate overloaded feeders and/or balance the demand of power with the available power supply.

The OPF objective is the minimization of the operating and maintenance costs associated with the generation of power for a given scenario $\vec{\vartheta}$ of duration t^h. Equation (12.5b) correspond to the power balance equation at node i, whereas Equations (12.5c) and (12.5d) are the bounds of the power generation, Equations (12.5e) and (12.5f) account for the technical limits of the feeders.

The available power in the distribution network is a function of the configuration Ξ and the mechanical states of the power sources:

$$P_{Ga_{i,J}}^{\vec{\vartheta}} = \xi_{i,j} mc_{i,j}^{\vec{\vartheta}} G_{i,j}^{\vec{\vartheta}}, \tag{12.6}$$

where, $G_{i,j}^{\vec{\vartheta}}$ represents the unitary power output and depends on the type of power source, i.e.

$$G_{i,j}^{\vec{\vartheta}} = \begin{cases} P_{i,j}^{ms}\left(mc_{i,j}^{\vec{\vartheta}}\right) & j \in M \\ P_{i,j}^{pv}\left(s_i^{\vec{\vartheta}}, mc_{i,j}^{\vec{\vartheta}}\right) & j \in P_V \\ P_{i,j}^{w}\left(ws_i^{\vec{\vartheta}}, mc_{i,j}^{\vec{\vartheta}}\right) & j \in w, \forall i \in N. \\ P_{i,j}^{ev}\left(op\left(td^{\vec{\vartheta}}\right), t_{Rop}^{\vec{\vartheta}}, mc_{i,j}^{\vec{\vartheta}}\right) & j \in \varepsilon_V \\ P_{i,j}^{st}\left(Q^{st\vec{\vartheta}}, mc_{i,j}^{\vec{\vartheta}}\right) & j \in S_T \end{cases} \tag{12.7}$$

12.1.3 Performance Indicators

Given a set γ of n^S sampled operational scenarios $\vec{\vartheta}_\ell$ and $\ell \in \{1,\ldots,n^S\}$, the OPF is solved for each scenario $\vec{\vartheta}_\ell \in \gamma$, giving in output the respective values of ENS and global cost.

ENS is a common index for reliability evaluation in power systems [3]. In the present work, this is obtained directly from the OPF output in the form of the aggregation of all-nodal load shedding per scenario $\vec{\vartheta}_\ell$:

$$ENS^{\vec{\vartheta}_\ell} = \frac{\sum_{i \in \mathcal{N}} LS_i^{\vec{\vartheta}_\ell}}{t^h}, \forall \vec{\vartheta}_\ell \in \gamma, \tag{12.8}$$

$$\overrightarrow{ENS}^{\gamma} = \left[ENS^{\vec{\vartheta}_1}, \ldots, ENS^{\vec{\vartheta}_\ell}, \ldots, ENS^{\vec{\vartheta}_{n^S}} \right]. \tag{12.9}$$

The global cost C_g of the distribution network is formed by two terms: fixed costs and variable costs. The former term includes those costs paid at the beginning of the operation after the installation of the DG (conception of $\Xi^{\mathcal{D}}$). The variable term refers to the operating and maintenance costs. These costs are dependent on the power generation and supply, which are a direct output of the OPF in Equation (12.5a). In addition, this term considers revenues associated to the renewable sources incentives as well as energy prices. Thereby, the global cost function for a scenario $\vec{\vartheta}_\ell$ is given by:

$$C_g^{\vec{\vartheta}_\ell} = \frac{\sum_{i \in \mathcal{N}} \sum_{j \in \mathcal{D}} \xi_{i,j} \left(C_{inv_j} + C_{O\&M_j^f} \right)}{t^h} \times t^S + C_{O\&M}^{net\vec{\vartheta}_\ell} - (inc + ep) \times \sum_{i \in \mathcal{N}} \sum_{j \in \mathcal{D}} P_{Gu_{i,j}}^{\vec{\vartheta}_\ell} \times t^S, \forall \vec{\vartheta}_\ell \in \gamma, \tag{12.10}$$

$$\overrightarrow{C_g}^{\gamma} = \left[C_g^{\vec{\vartheta}_1}, \ldots, C_g^{\vec{\vartheta}_\ell}, \ldots, C_g^{\vec{\vartheta}_{n^S}} \right]. \tag{12.11}$$

where

C_{inv_j}	investment cost of the DG technology j $[\$]$
$C_{O\&M_j^f}$	operating and maintenance fixed costs of the DG technology j $[\$]$
t^S	horizon of analysis $[h]$
inc	incentive for generation from renewable sources $[\$/kWh]$
ep	energy price $[\$/kWh]$
$C_g^{\vec{\vartheta}_\ell}$	global cost $[\$]$

The proposed MOO framework introduces CVaR as a coherent measure of the risk associated to the functions of interest. This risk measurement allows evaluating how "risky" is the selection of a determined value of expected losses. We consider a fixed configuration of the distribution network $\{\Xi, \mathcal{F}\}$ including the integration of DG units as a "portfolio." The assessed $\overrightarrow{ENS}^{\gamma}$ and $\overrightarrow{C_g}^{\gamma}$, found from the MCS-OPF to the set of scenarios γ, can be treated as estimations of the probability of the "losses." In this sense, if the decisions are intended to be taken based on the expectations of $\overrightarrow{ENS}^{\gamma}$ and $\overrightarrow{C_g}^{\gamma}$, then the $CVaR\left(\overrightarrow{ENS}^{\gamma}\right)$ and $CVaR\left(\overrightarrow{C_g}^{\gamma}\right)$ will represent the risk associated to these expectations.

As shown in Figure 12.2A, for a discrete approximation of the probability of the losses, given a confidence level or α-percentile, the value-at-risk VaR_α represents the smallest value of *losses* for which the probability that the losses do not exceed the value of VaR_α is greater than or equal to α. Thus, from the cumulative distribution function (cdf) *F(losses)*, it is possible to construct the α-tail cdf F_α*(losses)* for the losses, such as (Figure 12.2B): the α-tail cdf represents the risk "beyond the VaR" and its mean value corresponds to the $CVaR_\alpha$.

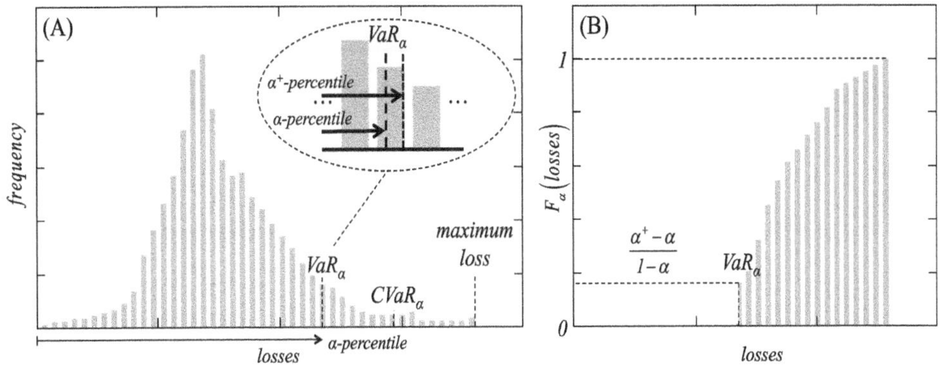

Figure 12.2 Graphical representation of the *CVaR* concept [4].

12.1.4 MOO Problem Formulation

The MOO problem consists of the two objective functions measuring the C_g and ENS and the associated risks. Specifically, their expected values and their *CVaR* values are combined, weighted by a factor of $\beta \in [0,1]$, which allows modulating the expected performance of the distribution network and its associated risk. Considering a set of randomly generated scenarios γ, the optimization problem is formulated as follows:

- Objective functions:

$$\min f_1 = \beta \times EC_g^\gamma + (1-\beta) \times CVaR_\alpha\left(\overrightarrow{C_g}^\gamma\right) \tag{12.12a}$$

$$\min f_2 = \beta \times EENS^\gamma + (1-\beta) \times CVaR_\alpha(\overrightarrow{ENS}^\gamma) \tag{12.12b}$$

- Constraints:

$$\xi_{i,j} = \begin{cases} 1 \, if \, \zeta DG \text{ technology j are allocated at node i} \\ 0 \; otherwise \end{cases} \quad \forall i \in \mathcal{N}, j \in \mathcal{D}, \zeta \in Z^+ \tag{12.12c}$$

$$\sum_{i \in \mathcal{N}} \sum_{j \in \mathcal{D}} \xi_{i,j}\left(Cinv_j + C_{O\&m_j^f}\right) \le BGT \tag{12.12d}$$

$$\sum_{i \in \mathcal{N}} \xi_{i,j} \le \tau_j, \forall j \in \mathcal{D} \tag{12.12e}$$

$OPF(\Xi, \mathcal{F}, \gamma)$ in Equations (12.4a)–(12.4f)

where EC_g and expected energy not supplied (EENS) denote the expected values of EC_g and ENS, respectively.

The meaning of each constraint is

Equation (12.11c)	the decision variable $\xi_{i,j}$ is a positive integer number
Equation (12.11d)	the total costs of investment, and fixed operation and maintenance of the DG units must be less than or equal to the available budget BGT
Equation (12.11e)	the total number of DG units of each technology j to allocate must be less than or equal to the maximum number of units τ_j available to be integrated
Equations (12.4a)–(12.4f)	all the equations of OPF must be satisfied for all scenarios in Υ

12.1.5 Solution Approach and Case Study Results

The combinatorial MOO problem under uncertainties is solved by the NSGA-II algorithm presented in Chapter 10. In this approach, the evaluation of the objective functions

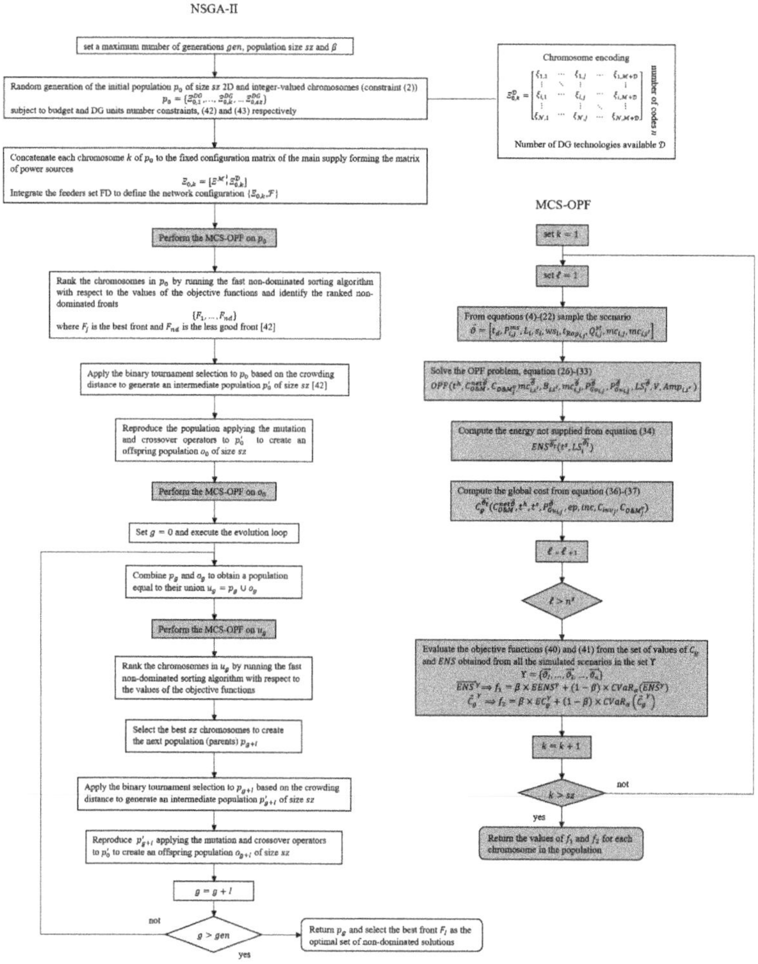

Figure 12.3 Flow chart of NSGA-II MCS-OPF MOO framework.

is performed by the developed MCS-OPF. The searching process of the overall NSGA-II MCS-OPF is summarized as shown in Figure 12.3.

As mentioned in Section 12.1.1, the testbed is modified from the IEEE 13 nodes test feeder (shown in Figure 12.1). The details about the characteristics of the components in the DG system can be found in [4]. The Pareto fronts, resulting from the MOO realizations for the different values of β, are presented in Figure 12.4. Each set of solutions corresponds to the "last-generation" population of the GA and the non-dominated solutions are presented in bold markers.

In Figure 12.5, the performance of the distribution network referring to the ENS and the C_g is improved for any realization of the MOO; if compared to the only MS case, it will show the gain in reliability of power supply and the economic benefits obtained by purchasing power from the different renewable DG sources. On the other hand, it is possible to infer that, in general, for lower values of the weight parameter β, the mean values of C_g is higher. This is expected, given that for the definition of the objective functions, when β tends to 0, the MOO tends to minimize the CVaR. (We skipped much of the analyses and discussions about the results because the intention of this chapter is only to illustrate through case studies the way of utilizing the methods and tools presented in previous Chapters. For the details about this case study, please refer to [4].)

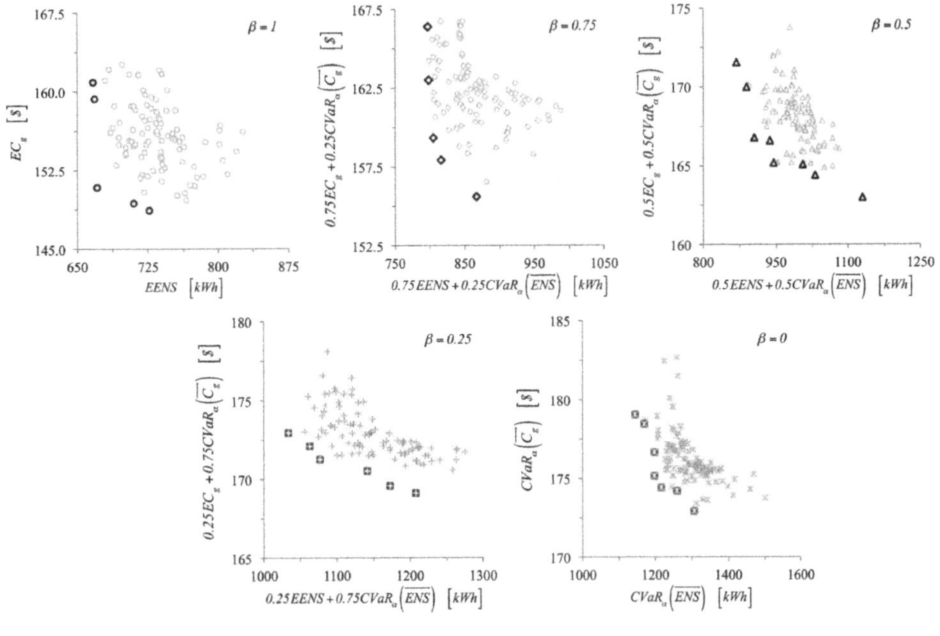

Figure 12.4 Pareto fronts for different values of β [4].

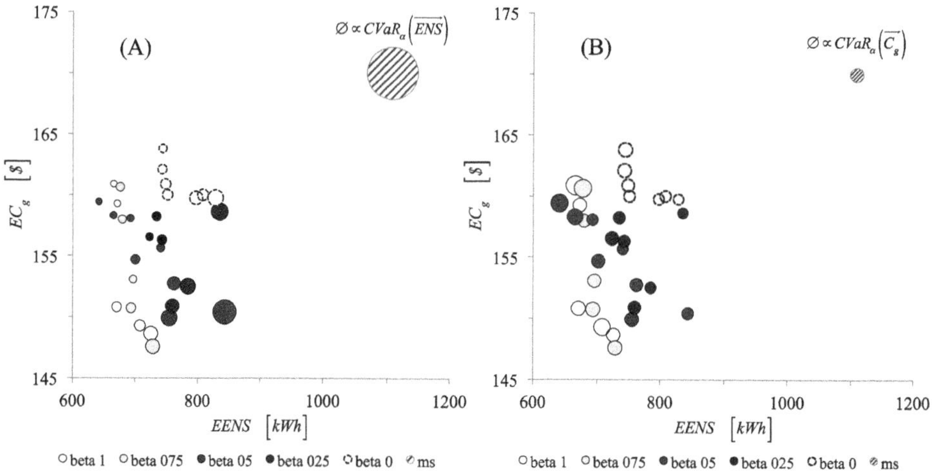

Figure 12.5 EENS v/s ECg [4].

12.2 Redundancy Allocation for Binary-State Series-Parallel Systems (BSSPSs) under Epistemic Uncertainty

In this section, we consider the redundancy allocation problem (RAP) with uncertain data in a binary-state series-parallel system (BSSPS). We assume the states of components and system are binary, the states of individual components are statistically independent, and the redundancy strategies in all subsystems are active.

12.2.1 Problem Description

We consider a BSSPS with $|I|$ subsystems connected in series. In each subsystem $i \in I$, the decision variables are the redundancy levels x_i of the components to be placed in parallel. The system cost is minimized under a system reliability requirement. The RAP model is given by

$$\min \sum_{i \in I} \tilde{c}_i x_i \tag{12.13a}$$

$$\text{s.t.} \prod_{i \in I} \left(1 - \left(1 - r_i\right)^{x_i}\right) \geq R_0 \tag{12.13b}$$

$$L_i \leq x_i \leq U_i, \forall i \in I \tag{12.13c}$$

$$x_i \in \{0,1\}, \forall i \in I \tag{12.13d}$$

where Equation (12.13b) represents the requirement that the system reliability should be larger than R_0, Equations (12.13c) shows the range of redundancy levels in each subsystem.

The reliability function can be linearized with binary variables [5]. Therefore, the model in Equations (12.13a)–(12.13d) can be reformulated as the following integer programming problem:

$$\min \sum_{i \in I} \tilde{c}_i x_i \tag{12.14a}$$

$$\text{s.t.} \sum_{i \in I} \sum_{k \in K_i} \chi_{ik} \ln\left[1-(1-r_i)^{L_i+k}\right] \geq \ln R_0 \tag{12.14b}$$

$$\sum_{k \in K_i} \chi_{ik} = 1, \forall i \in I \tag{12.14c}$$

$$x_i = \sum_{k \in K_i} k\chi_{ik} + L_i \tag{12.14d}$$

$$\chi_{ik} \in \{0,1\}, \forall i \in I, k \in K_i \tag{12.14e}$$

where the binary variable χ_{ik} denotes whether $x_i - L_i$ equals to $k \in K_i$ and $K_i = \{0,1,\ldots,U_i - L_i\}$. Therefore, the redundancy level x_i can be replaced by $\sum_{k \in K_i} k\chi_{ik} + L_i$ with χ_{ik}. The constraint in Equation (12.13c) ensures that only one redundancy level $L_i + k$ for $k \in K_i$ is selected for the redundancy level x_i. The reliability function is log-transformed to the linear forms

$\sum_{i \in I} \ln\left(1-(1-r_i)^{x_i}\right) = \sum_{i \in I} \sum_{k \in K_i} \chi_{ik} \ln\left(1-(1-r_i)^{L_i+k}\right)$ as in Equation (12.14b).

12.2.2 Robust Model

We consider that the parameters of cost $\tilde{c}$ are uncertain. In practice, it is reasonable to estimate the mean (nominal) values c_i and variation ranges $\hat{c}_i$ of these parameters for all possible component types. We assume all cost parameters $\tilde{c}_i$, $i \in I$ are mutually independent, symmetric, and bounded, which take values in $[c_i - \hat{c}_i, c_i + \hat{c}_i]$. Therefore, the robust model with polyhedral uncertainty set is used to handle the RAP with uncertain data. The uncertainty set of $\tilde{c}$ is denoted as follows:

$$C := \left\{c \in R_+^{|I|} : \tilde{c}_i \in \left[c_i - \hat{c}_i, c_i + \hat{c}_i\right], \forall i \in I\right\}$$

For the constraints with uncertain data, the robust formulation of objective in Equation (12.14a) with $\tilde{c} \in C$ is given by

$$\min_x \max_{\tilde{c} \in C} \sum_{i \in I} \tilde{c}_i x_i \tag{12.15}$$

The general uncertainty sets considered in robust models include polyhedral uncertainty set [6] and ellipsoidal uncertainty set [7]. In this RAP, we use the cardinality constrained uncertainty proposed in [8] to deal with the polyhedral uncertainty on $\tilde{c}$. The protection level $\Gamma^c \in [0,|I|]$ is introduced to control the robustness of the model on Equation (12.15). The cardinality constrained robust representation of Equation (12.15) is given by

$$\sum_{i\in I} c_i x_i + \max_{\{S\cup\{s\}:S\subseteq I,|S|=\lfloor\Gamma^c\rfloor,s\in I\setminus S\}} \left\{\sum_{i\in S} \hat{c}_i x_i + \left(\Gamma^c - \lfloor\Gamma^c\rfloor\right)\hat{c}_s x_s\right\} \tag{12.16}$$

i.e. up to $\lfloor\Gamma^c\rfloor$ of all uncertain parameters are allowed to change by $\hat{c}_i$ and one parameter $\hat{c}_{s_i}$ is allowed to change by $\left(\Gamma^c - \lfloor\Gamma^c\rfloor\right)\hat{c}_s$. Finally, the robust optimization model for the BSSPS RAP (12.14a)–(12.14e) is as follows:

$$\min \sum_{i\in I} c_i x_i + \max_{\{S\cup\{s\}:S\subseteq I,|S|=\lfloor\Gamma^c\rfloor,s\in I\setminus S\}} \left\{\sum_{i\in S} \hat{c}_i x_i + \left(\Gamma^c - \lfloor\Gamma^c\rfloor\right)\hat{c}_s x_s\right\} \tag{12.17a}$$

$$\text{s.t.}(12.13\text{b}) - (12.13\text{e}) \tag{12.17b}$$

Given the robust model in Equations(12.17a)–(12.17b) in the BSSPS RAP, we present the tractable formulation for it in this section. The formulationin Equation (12.17a) is intractable because the combinations of $\left\{S\cup\{s\}: S\subseteq I, |S| = \lfloor\Gamma^c\rfloor, s\in I\setminus S\right\}$ compared under the operation "max" are exponential. These semi-infinite formulations can be transformed into linear formulations through a duality argument.

Proposition 1 Given a decision x, the semi-infinite formulation in Equation (12.17a) is equivalent to the following program:

$$\min \sum_{i\in I} c_i x_i + \xi^c \Gamma^c + \sum_{i\in I} \rho_i^c \tag{12.18a}$$

$$\xi^c + \rho_i^c \geq \hat{c}_i x_i, \forall i \in I \tag{12.18b}$$

$$\rho_i^c \geq 0, \forall i \in I \tag{12.18c}$$

$$\xi^c \geq 0 \tag{12.18d}$$

$$x_i \in Z_+, \ \forall i \in I \tag{12.18e}$$

where ξ^c, ρ_i^c are auxiliary variables.

Proof The proof of Proposition 1 is given in Section 11.3.3.
Therefore, the tractable formulation of the model in Equations (12.17a)–(12.17b) is given by

$$\min \sum_{i\in I} c_i x_i + \xi^c \Gamma^c + \sum_{i\in I} \rho_i^c$$
$$(12.17\text{b}) - (12.17\text{e}),\ (12.13\text{b}) - (12.13\text{e})$$

12.2.3 Experiment

We consider a BSSPS of 10 subsystems and a reliability requirement larger than 0.9. The nominal cost parameters c_i are uniformly generated in the range $[10,15]$ in arbitrary units of cost and $\tilde{c}_i$ is set as $c_i/2$. The reliability of each component in subsystem $i \in I$ is uniformly generated from $[0.85,0.90]$. The bounds of the redundancy level for subsystem i is $L_i = 1$ and $U_i = 5$, for $i \in I$.

To illustrate the performance of protection level Γ^c on the robust model, we vary Γ^c in the set $\{0,1,\ldots,10\}$. Given a protection level Γ^c, the associated robust solution and objective are represented as $x^*(\Gamma^c)$ and $C^*(\Gamma^c)$, respectively. To explore the robustness of the solutions $x^*(\Gamma^c)$, 100,000 samples for the cost parameters $\tilde{c}^s$ are simulated from the uniform distribution, normal distribution, or triangle distribution on C. The objective value $C^s(\Gamma^c)$ of solution $x^*(\Gamma^c)$ under the sample $\tilde{c}^s$ is calculated by $(\tilde{c}^s)^{\mathrm{T}} x^*(\Gamma^c)$. Then the violation probability of the robust solution is represented by the frequency that $C^*(\Gamma^c)$ is less than $C^s(\Gamma^c)$.

Figure 12.6 shows the violation probabilities for different Γ^c. The violation probability drops sharply with the increase in the protection level. Actually, when $\Gamma^c = 2$, the violation probability is 0.004 and when Γ^c=4, the violation probability has become smaller than 10^{-5}.

Figure 12.7 shows the change percentage in objective values given by

$$\frac{C^*(\Gamma^c) - C^*(0)}{C^*(0)} \times 100\%$$

considering different levels of robustness. As we increase the protection level Γ^c and the optimal value increases with Γ^c, the robust solution becomes more conservative. According to Figures 12.6 and 12.7, we observe that by allowing the cost to increase by 0.08, we can make the probability of constraint violation less than 0.04. In addition, by

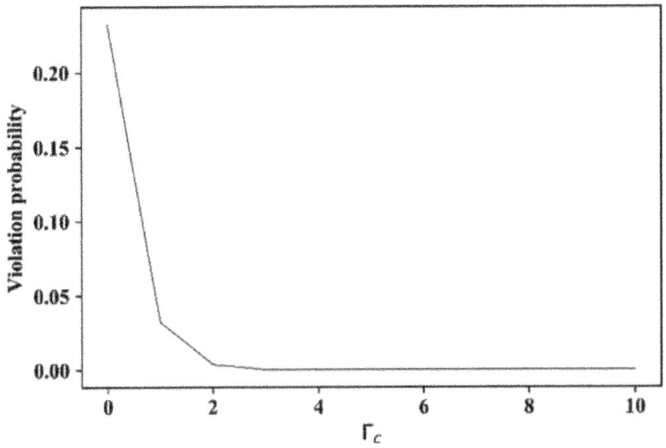

Figure 12.6 Violation probabilities for different Γ^c.

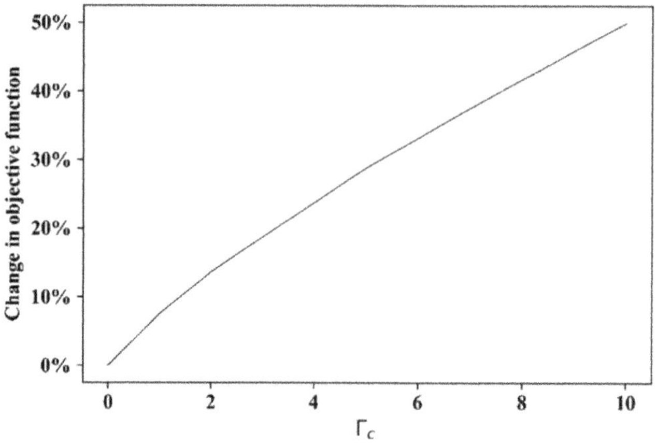

Figure 12.7 Change percentage in objective function for varying Γ^c.

allowing the cost to increase by 0.24, the violation probability is less than 10^{-5}. Therefore, we can sacrifice a relatively small increment in the objective value to greatly reduce the violation probability.

References

1 Rezaei, F., Najafi, A.A., and Ramezanian, R. (2020). Mean-conditional value at risk model for the stochastic project scheduling problem. *Computers Industrial Engineering* 142: (106356).

2 Schneider, K.P. et al. (2017). Analytic considerations and design basis for the IEEE distribution test feeders. *IEEE Transactions on Power Systems* 33 (3): 3181–3188.

3 Azaron, A., Perkgoz, C., Katagiri, H., Kato, K., and Sakawa, M. (May 2009). Multi-objective reliability optimization for dissimilar-unit cold-standby systems using a genetic algorithm. *Computers & Operations Research* 36 (5): 1562–1571. doi:10.1016/j.cor.2008.02.017.

4 Mena, R., Hennebel, M., Li, Y.F., Ruiz, C., and Zio, E. (Sep 2014). A risk-based simulation and multi-objective optimization framework for the integration of distributed renewable generation and storage. *Renewable & Sustainable Energy Reviews* 37: 778–793. doi:10.1016/j.rser.2014.05.046.

5 Feizollahi, M.J. and Modarres, M. (2012). The robust deviation redundancy allocation problem with interval component reliabilities. *IEEE Transactions on Reliability* 61 (4): 957–965.

6 Soyster, A.L. (1973). Convex programming with set-inclusive constraints and applications to inexact linear programming. *Operations Research* 21 (5): 1154–1157.

7 Ben-Tal, A. and Nemirovski, A. (2000). Robust solutions of linear programming problems contaminated with uncertain data. *Mathematical Programming* 88 (3): 411–424.

8 Bertsimas, D. and Sim, M. (2004). The price of robustness. *Operations Research* 52 (1): 35–53.

Index

Reliability Analysis, Safety Assessment and Optimization: Methods and Applications in Energy Systems and Other Applications, First Edition. Enrico Zio and Yan-Fu Li.